MITOCHONDRIAL DYSFUNCTION IN BIOLOGICAL AGING AND CANCER

BIO HACKS FOR HEALTHY AGING AND LONGEVITY WITH SUPPLEMENTS:
COQ10, NITRIC-OXIDE, L-GLUTATHIONE, NMN, TAURINE, MELATONIN, NIACIN, METHYL-COBALAMIN, UROLITHIN-A, GLYNAC, NIACIN, NAD+

S. PRAKASH

ISBN

Hardcase 979-8-89556-978-8
Paperback 979-8-89519-878-0

First Edition: September 2024
kaalhasthi@gmail.com

वक्रतुण्ड महाकाय सूर्यकोटि समप्रभ।
निर्विघ्नं कुरु मे देव सर्वकार्येषु सर्वदा

Table of Contents

Table of Contents

About the Book

This book is about Mitochondrial dysfunction and how it impacts the cellular aging of different cells, tissues, and organs.

This book caters to everyone, helping them understand the basic factors that age you faster and how to combat this. However, it can also be used as a ready reckoner by nutritionists, medical students, biochemists, biotechnologists, fitness freaks, future physicians, and health enthusiasts.

The author has backed all the articles on supplements, vitamins, or certain molecules from the online resources available on PubMed (PubMed (nih.gov)), the National Library of Medicine—National Institutes of Health (https://www.nlm.nih.gov/), Frontiers in Medicine, (ScienceDirect.com | Science, health and medical journals, full text articles and books). (Cleveland Clinic: Every Life Deserves World Class Care) (Top-ranked Hospital in the Nation - Mayo Clinic) MDPI - Publisher of Open Access Journals and other online resources mentioned in all the articles of every chapter.

Mitochondrial diseases/dysfunction are a group of metabolic disorders (Obesity, Diabetes Type 2 (T2DM), Fatty Liver and Inflammatory Bowel Diseases, Alzheimer's disease, Parkinson's disease etc.).

Mitochondrial dysfunction can be caused by various factors such as **Inflammation, Oxidative Stress, Insulin Resistance, Multiple sclerosis, Muscular Dystrophy and Cancer** which can lead to the following diseases: **KIDNEY FAILURE, LIVER FAILURE, HEART ATTACK, PCOS, CANCER, DIABETES, OBESITY, PROSTATE ENLARGEMENT, FITS, INFERTILITY, ULCERATIVE COLITIS, LUNGS FAILURE, BLINDNESS, ERECTILE DYSFUNCTION, HORMONAL IMBALANCE, OVARIAN CYST, UTERUS FIBROIDS, AUTOIMMUNE DISEASE, INSOMNIA.**

The author has done his research by going through online medical resources of bio hacks to prevent cancer and delay cellular aging by following supplements: **CoQ10, Nitric-Oxide, L-glutathione, NMN, Taurine, Melatonin, NIACIN, Methyl-Cobalamin, Urolithin-A, GlyNAC, NIACIN, NAD+, SIRT6, Amino Acid, Omega Fatty 3 Acids, Vitamin D3 and Magnesium etc.**

The author strongly recommends that you consult your physician before beginning any exercise program. The author is not a licensed healthcare provider, and he has no expertise in diagnosing, examining, or treating medical conditions of any kind or determining the effect of any specific exercise on a medical condition.

The information presented in this book is strictly for educational purposes. The author urges you **not to** engage in self-diagnosis or treatment based solely on the knowledge acquired from this book. You must consult a medical practitioner for any health concerns, as this book does not claim to provide any medical solutions for any health issues.

The content of this book is for informational purposes only and is not intended to diagnose, treat, cure, or prevent any condition or disease. You understand this book is not intended as a substitute for consultation with a licensed practitioner. Please consult your physician or healthcare specialist regarding the suggestions and recommendations made in this book. The use of this book implies your acceptance of this disclaimer.

The author advises you to take full responsibility for your safety and not to take any risks by engaging in self-treating, as this book is for educational purposes only.

Disclaimer

Before we proceed, we want to remind you that **the articles on different supplements in the book are not a substitute for professional advice.**

The author is not a doctor or dietician. However, as a researcher in Medical Astrology, he has done extensive reading over the past few years on Mitochondrial Dysfunction and how it affects the cellular aging of cells, tissues, and organs, which has inspired him to write on this subject.

This book presents information strictly for educational purposes. It's tailored for those interested in the latest "health and wellness" research.

The author urges you not to engage in self-diagnosis or treatment based solely on the knowledge acquired from this book. You must consult a medical practitioner for any health concerns, as this book does not claim to provide any medical solutions for any health issues.

The content of this book is for informational purposes only and is not intended to diagnose, treat, cure, or prevent any condition or disease. You understand this book is not intended as a substitute for consultation with a licensed practitioner. Please consult your physician or healthcare specialist regarding the suggestions and recommendations made in this book.

This publication is meant as a source of valuable information for the reader; however, it is not meant as a substitute for direct expert assistance. If such assistance is required, the services of a competent professional should be sought.

My Astrological Predictions

My Predictions (WhatsApp chat transcript) are on my website and on my Facebook page.

Website

Https://www.kaalhasthiastrologer.com

Facebook
facebook.com/kaalhasthi.astro

YouTube channel
CosmicKrishna
https://www.youtube.com/channel/
UCFObfsKxwkMV-TRk9fbDI2g

About the Author

A Science graduate and project management certified professional in the IT industry. He spends his time studying Vedanta/Upanishads/Darshan Shastra and Puranas. In addition, he does research in Astrology and Medical Science. He also teaches Astrology and writes books on the above subjects.

S. Prakash has done good research on Medical Astrology, DNA Astrology, and Past Life curses and how to curb them as per Vedic Literature from Vedanta/Upanishad and Puranas.

S.Prakash – Philosopher, Mythologist, Author and Astrologer

Life Transformation Coach through Past Life Pending Karma

We have the lock in the form of luck, but sometimes, we cannot find the right key to unlock it.

Dedication

This book could not have been completed without expressing my gratitude to my teacher and my Guru, who has been an inspiration in my life. Therefore, I would like to dedicate this book to My Guru, Late Shree J.N Sharma Ji.

My book is published with Guruji's blessings, which I seek at his lotus feet.

Acknowledgement

I want to acknowledge my father, **Shri Satya Prakash Gupta** and my mother, **Mrs. Vijay Gupta**, for their blessings, guidance and support.

I also want to acknowledge my Grandparents, the **Late Shri Devaki Nandan Varshney** and the **Late Smt. Ramkali Devi** for their eternal blessings.

I seek to acknowledge my wife's grandmother (**Smt. Maya Devi**), who reads my books and inspires me to contribute to Astrology by writing more books.

My wife, Mrs. **Lavi Gupta** (Shivani), and My daughter, Gauri (**Gunika Gupta**), have motivated me to pursue my interest of authorship.

I also want to mention the name of my brother-in-laws, **Hemant Kumar Gupta and Gourav Gupta who are in the medical field and inspired me to write a book on medical science.** I cannot thank enough my siblings, **Ajay Gupta (brother) and Akanksha Gupta (sister),** who has been a great support system throughout my life.

Previous Books written by Author

The Author has written below books:

1. Snapshot Prediction through Yogini Dasha.
2. Unlock Pitra Dosh with Lal Kitab Pending Karma through Shreemad Bhagwat Puran.
3. Unlock Pending Karma and its correction.
4. पिछले जीवन के अधूरे कर्म और उसके सुधार को अनलॉक करें.
5. Unlock Purva Punya and Paap from the stories of 27 Nakshatra.
6. Unlock Luck and Wealth with keys of Dharma.
7. Unlock Marital Curse through Rashi Tulya Navamsha.
8. Autism in Medical Astrology.
9. DNA Astrology of Wealth through Nakshatra with Bhrighu Nandi Nadi.
10. Business Yoga with Bhava Bala and Apokilam Houses with D10 Chart.
11. Prashna Kundali and Saptarishis in Manvantara and Yugas - Plan of Brahma in Creation of Universe.
12. Destiny Vs Karma (Free-Will)Through DNA Astrology from Past Life Karma - What you Sow, Not always reaped, as God has a better Plan -
Inspired by Bhagwad Geeta and Mythologies from Puran
13. भृगु नंदी नाडी शुक्र गृह के डीएनए से - कैसे देखें धन योग (अष्टलक्ष्मी योग), नक्षत्र से, इंदु लग्न से और शुक्र लग्न के साथ.
14. लाल किताब से पितृ दोष और पूर्व जन्म के पितृ ऋण के कर्म का सुधार-डिकोड पेंडिंग कर्म, श्रीमद् भागवत पुराण से, और अपने जीवन के दोषो से मुक्ति पाने की कुंजी.

The Author has his Youtube channel by the following name: **CosmicKrishna**

On this channel, you will find many astrological videos and the glimpse of all the courses.

S. Prakash

S. Prakash has written and published the book **"Unlock Pending Karma and its Correction"** in English language.

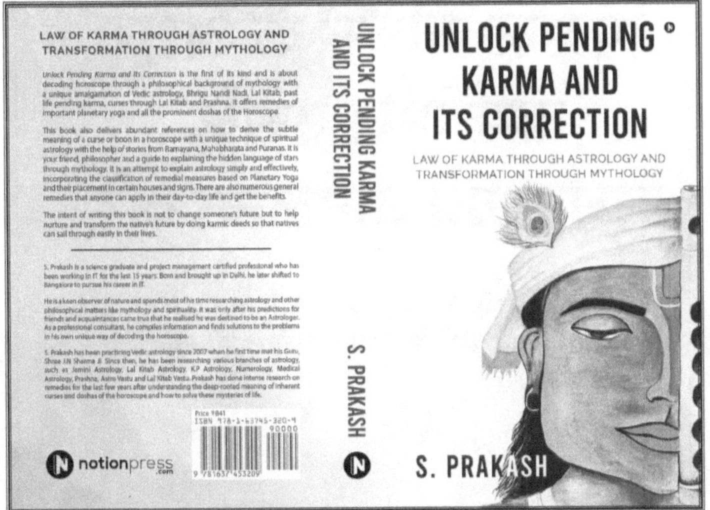

Law of Karma through Astrology and Transformation through Mythology

Book is available across the globe on all the platforms and have been best-selling on Amazon.

S.Prakash has written and published the book "**पिछले जीवन के अधूरे कर्म और उसके सुधार को अनलॉक करें**" in Hindi language.

S. Prakash has written and published his book - "Unlock Purva Punya and Paap from the Stories of 27 Nakshatras" in English language.

Curses through Medical Astrology

S. Prakash has written – **"Autism in Medical Astrology with Arudha Lagna of Jaimini Jyotish and Chandra Lagna of Parashar Vedic Jyotish"** in English language.

Decoding Past Life Sin through Nakshatra

S. Prakash has written and published the book:

Prashna Kundali and Saptarishis in Manvantara and Yugas

Plan of Brahma in Creation of Universe

S. Prakash has written and published the below book:

S. Prakash has written and published the below book:

S. Prakash has written and published the below book:

S. Prakash has written and published the below book:

S. Prakash has written and published the below book:

S. Prakash has written and published the below book:

S. Prakash has written and published the below book:

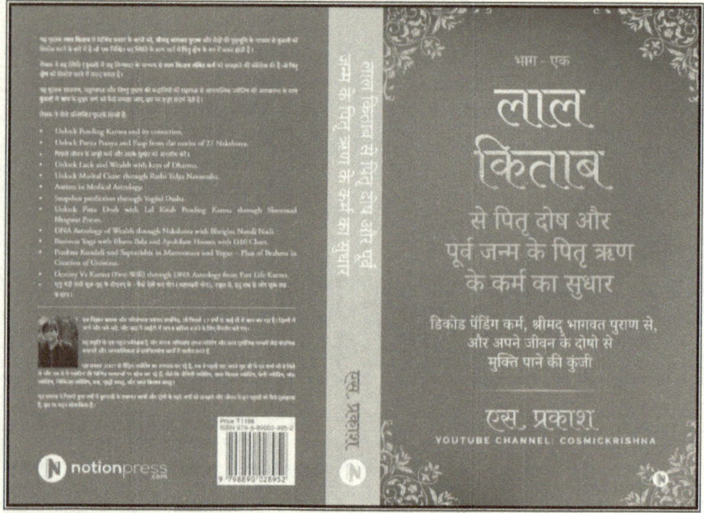

S. Prakash has written and published the below book:

Introduction

Metabolic Syndrome and Mitochondrial dysfunction

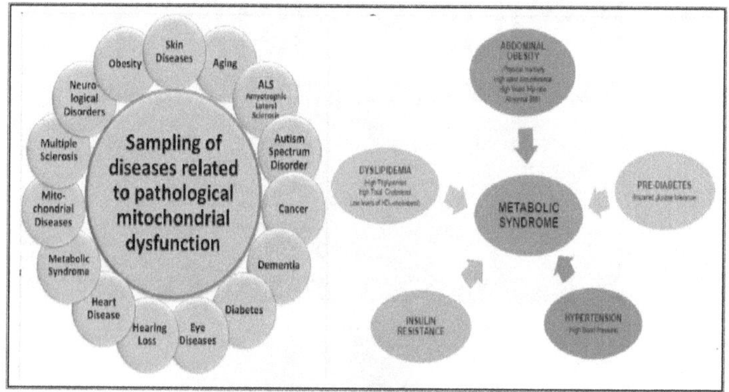

Pictures Courtesy from –

https://www.researchgate.net/figure/Diseases-related-to-pathological-mitochondrial-dysfunction_fig2_361930961

https://europepmc.org/article/med/27836629

Metabolic syndrome, a cluster of conditions, is closely linked to obesity, high blood pressure, high triglycerides/LDL, low levels of HLD cholesterol, obesity/dyslipidaemia, hypercoagulability, and insulin resistance. It's important to understand that metabolic syndrome is not a disease itself but a group of risk factors that occur together, increasing the risk of heart disease, stroke, and type 2 diabetes.

The Pancreas, a vital organ, produces the hormone insulin. Insulin is key to maintaining normal blood sugar levels and plays a crucial role in allowing glucose from the bloodstream to enter the body's cells, where it is used for energy. Understanding this process is essential for managing and preventing health issues related to insulin resistance.

However, when the body cannot take the glucose into its cells, extra sugar remains in the bloodstream, and blood sugar levels are not controlled properly, resulting in type 2 diabetes. This is when the Pancreas becomes exhausted and unable to control the blood sugar levels, or the body becomes immune to insulin, which is called **Insulin resistance**. Insulin resistance, a condition that demands immediate attention, if left unmanaged, can lead to serious health issues such as type 2 diabetes and heart disease.

Having metabolic syndrome of Insulin Resistance can increase the risk of developing Type 2 diabetes, Heart and blood vessel

disease, Hypertension, PCOS, Fatty Liver, and Cancer.

The digestive system breaks down the foods into sugar. Insulin is a hormone made by your Pancreas that lets sugar enter your cells to be used as fuel. In insulin-resistance conditions, cells don't respond typically to insulin, and glucose can't enter the cells so quickly. As a result, your blood sugar levels rise, and Pancreas releases more insulin to lower your blood sugar. Insulin resistance is the inability of cells to respond to insulin, which is the main feature of metabolic syndrome that leads to high blood sugar and type 2 diabetes. In simpler terms, it's like the cells in your body are not listening to the insulin's instructions to let glucose in, which can lead to high blood sugar levels.

Metabolic syndrome (MetS) is a group of many metabolic abnormalities, including hypertension, hyperglycemia, abdominal obesity, and dyslipidemia represented by low HDL cholesterol and hypertriglyceridemia. These conditions occur together and increase the risk of type 2 diabetes and cardiovascular diseases.

Metabolic syndrome represents a constellation of many risk factors, such as obesity, higher blood pressure, abnormal levels of triglycerides and cholesterol, etc., which are involved in the higher prevalence of non-communicable diseases. Altered mitochondrial functioning has been implicated in the pathophysiology of type 2 diabetes, obesity, dyslipidaemia, and cardiovascular diseases.

The imbalance between energy production and its utilization may lead to defective cell metabolism, which is considered one of the main culprits of metabolic syndrome. Increased glucose levels enhance ROS overproduction, leading to morphological changes in mitochondria.

Inhibition of the insulin signalling pathway causes the accumulation of lipids and free fatty acids (FFA), which contributes to metabolic disorders. Also, aging, altered mitochondrial biogenesis, decreased antioxidant defence capacity, and genetic factors cause insulin resistance, the primary cause of many metabolic diseases.

Insulin resistance (IR) is characterized by the diminished capacity of the cells to respond to physiological insulin levels. Various risk factors, including aging, physical inactivity, abdominal obesity, and stress, contribute to the pathophysiology of insulin resistance.

Oxidative stress induced by an excess of ROS in mitochondria might also contribute to the development of insulin resistance. Few studies have shown that oxidative stress and mitochondrial dysfunction are involved in the pathophysiology of many metabolic disease states, such as insulin resistance, obesity, diabetes, and cardiovascular and neurodegenerative diseases.

However, it is still unclear whether only insulin resistance is the primary cause of mitochondrial dysfunction or vice versa. Nonetheless, many studies have proven the association of mitochondrial dysfunction with insulin resistance in various tissues.

In skeletal muscle, altered mitochondrial functions reduce ATP synthesis, and increased ROS generation leads to insulin resistance and obesity/diabetes.

The imbalance between energy production and its utilization may result in impaired cell metabolism, which is considered one of the main culprits of metabolic syndrome. Increased glucose levels enhance the overproduction of ROS, which leads to morphological changes in mitochondria.

Inhibition of the insulin signaling pathway causes the accumulation of lipids and free fatty acids (FFA), which contributes to insulin resistance and other metabolic disorders.

Oxidative stress and Mitochondrial Dysfunction

Oxidative stress refers to the imbalance between the production of ROS and antioxidants. The damaging effects of ROS are more potent than the compensatory effect of antioxidants in the cells. Mitochondrial dysfunction is defined as diminished mitochondrial biogenesis, altered membrane potential, decreased mitochondrial number, and altered activities of oxidative proteins due to the accumulation of ROS in cells and tissues. Regular metabolism of oxygen generates reactive oxygen species as a by-product.

A link between stroke and Mitochondrial Dysfunction

Stroke is the fourth leading cause of adult disability and mortality in the developing world, associated with social and economic problems. Oxidative stress is one of the major factors leading to cellular damage during ischemic brain injury.

Mitochondrial dysfunctions have been demonstrated as a critical player in the development of brain stroke, as evidenced by reduced ATP production, the starvation of glucose and oxygen to the tissues, and the influence on cell death pathways.

In experimental stroke models, the diminished supply of glucose and oxygen leads to impaired oxidative metabolism in brain tissue. The resultant oxygen-glucose deprivation causes ROS formation.

A link between obesity and Mitochondrial Dysfunction

Currently, obesity has become one of the major global health problems. It is one of the principal components of metabolic syndrome and is known to be a significant risk factor in the development of many metabolic disorders.

Although obesity is caused by the interaction of genetic and environmental factors, many studies have revealed its role in mitochondrial dysfunction. Recent studies have demonstrated the role of mitochondrial dysfunction in the pathogenesis of metabolic syndrome.

There is an overproduction of ROS in adipose tissues with altered activities of NADPH oxidase and antioxidative enzymes in obese mice. Intriguingly, abdominal obesity has been associated with defective mitochondrial biogenesis manifested by impaired mitochondrial dysfunction, oxidative metabolism, low mitochondrial gene expression, and reduced ATP generation in rodents and humans.

Mitochondrial dysfunction contributes to the severity of many pathological conditions, including skeletal muscle atrophy, diabetes, cardiovascular diseases, metabolic disorders, and neurodegenerative diseases. Recent studies have indicated that physical activity offers many benefits through improved insulin sensitivity and mitochondrial biogenesis in skeletal muscles in T2D patients.

Also, a significant increase in muscle mitochondrial respiration, mitochondrial content, oxidative enzyme activity, and mitochondrial density was observed in T2D patients after exercise. Exercise stimulates AMPK, activating PGC 1 by direct phosphorylation of threonine and serine residues.

Ref From –

https://www.ncbi.nlm.nih.gov/pmc/articles/ PMC5423868/

https://doi.org/10.1016/j.bbadis.2016.11.010

García-García, F., Monistrol-Mula, A., Cardellach, F., & Garrabou, G. (2020). Nutrition, Bioenergetics, and Metabolic Syndrome. Nutrients, 12(9), 2785.

Bhatti, J. S., Kumar, S., Vijayan, M., Bhatti, G. K., & Reddy, P. H. (2017). Therapeutic Strategies for Mitochondrial Dysfunction and Oxidative Stress in Age-Related Metabolic Disorders. Progress in Molecular Biology and Translational Science. https://doi.org/10.1016/bs.pmbts.2016.12.012

Robaglia, A., Cau, P., Bottini, J., & Seïte, R. (1989). Effects of isolation and high helium pressure on the nucleolus of sympathetic neurons in the rat superior cervical ganglion. Journal of the Autonomic Nervous System. https://doi.org/10.1016/0165-1838(89)90114-8

Metabolic Syndrome

Insulin Resistance

Insulin resistance is a condition in which the body is not able to use Insulin effectively and as a result cells become less responsive to Insulin, and then it leads to cardiovascular disease (CVD), Non-Alcoholic fatty liver, PCOS, and chronic kidney disease (CKD).

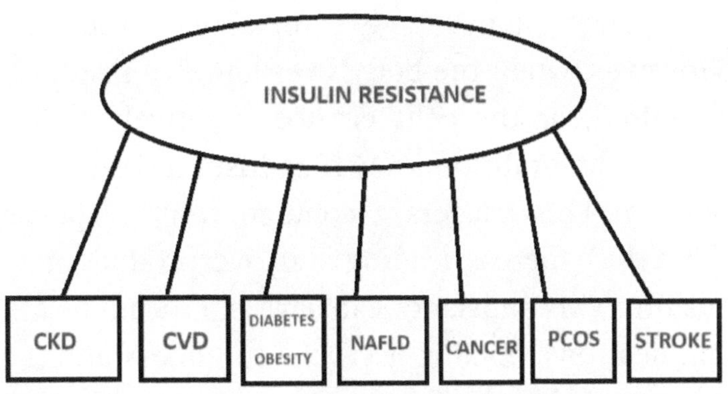

Insulin resistance is a condition where the cells in your liver, muscles don't respond effectively to insulin, and it can lead to serious implications for your health. Insulin, a hormone produced by your pancreas, is essential for managing your blood glucose. However, when you have insulin resistance, this process is disrupted. Your cells aren't able to effectively absorb glucose, leading to a buildup of blood glucose in your bloodstream.

Despite the cells becoming insulin-resistant, your pancreas continues to produce insulin. This persistence of insulin production is a key factor in the ongoing rise of blood glucose levels.

Insulin, a hormone that regulates the level of blood sugar (Glucose) level in the blood and aids in glucose transportation into the cells, is a key player in our body's metabolic processes. However, when the body's response to insulin is impaired, or the cells cannot effectively utilize insulin, a condition known as Insulin Resistance sets in. This triggers a compensatory response from the pancreas, leading to an overproduction of insulin. This imbalance can have significant health implications, making it crucial to understand the concept of **Insulin Resistance.**

Some of the common diseases caused by Insulin Resistance are:

Prediabetes

Type 2 diabetes

Polycystic ovarian syndrome (PCOS)

Microvascular disease (Retinopathy, Neuropathy and Nephropathy)

Macrovascular disease (Brain Stroke)

Macrovascular disease (PAD - Peripheral arterial disease)

Macrovascular disease (CAD - coronary artery disease)

Non-alcoholic fatty liver disease (NAFLD)

Obesity

Associated Symptoms are:

Hypertension

Hyperlipidemia

Inflammation

Endothelial dysfunction

Hyper Uricemia

Protein uricemia

Dyslipidemia

Hyper Glycemia

Hyper Coagulability

Obesity

Skin Tag

SNS activity

Insulin Resistance causes Fatty Liver (Non-Alcoholic or Alcoholic)

The liver has a wide range of functions such as detoxification and the production of bile to help with digestion. It also plays a large role in metabolism and producing albumin (protein), filtering blood and regulating glucose levels.

Fatty Liver is a condition in which fat builds up in the liver which is called as hepatic steatosis. There are of two types:

1. **NAFLD** - Nonalcoholic fatty liver disease

    Fatty Liver - This signifies that you have fat deposit in the liver, but you may not have any inflammation in your liver or no severe damage to your liver cells.

    Nonalcoholic steatohepatitis (NASH) - This is rather serious condition than fatty liver. NASH means that there is an inflammation in the liver.

2. **Alcoholic Fatty Liver**

    Fatty Liver refers to a condition wherein an excessive accumulation of fat deposit in the liver such as cholesterol and triglycerides.

NAFLD (Non-Alcoholic Fatty Liver Disease) – Fatty liver condition results when the body produces too much fat or does not metabolize fat effectively then this excess fat gets stored in the liver cells where it accumulates and cause fatty liver disease which could be due to Insulin Resistance.

Pancreas releases insulin which is dispersed into the bloodstream in response to the blood sugar level. **High Blood insulin level increase triglycerides which deposit fatty acids in the liver.**

Alcoholic fatty liver disease, also called alcoholic steatohepatitis leads to a buildup of fat inside your liver cells. This makes it harder for your liver to work. Steatotic (fatty) liver can happen in anyone who consumes a lot of alcohol.

Calories from alcohol are stored in the liver as fat. Liver fat makes liver cells more insulin resistant and can make your blood sugars higher over time. This may create the situation of Hyperglycemia, which means the too much sugar in your blood. This happens when your body has too little insulin or if your body can not use insulin properly which is called Insulin resistance which may contribute to the development of fatty liver (by impairing the ability of insulin to suppress lipolysis of breaking down of fats and lipids).

Insulin Resistance causes (CKD) Chronic Kidney Disease

The kidney is part of the Urinary System (the Human excretory system). It filters wastes (Urea, Uric acid) and excess fluids from the blood, which are then removed in the urine.

The heart pumps blood filled with oxygen through all parts of your body, including the kidney.

Without the heart, your kidneys would not have the oxygen-filled blood needed to do its many vital jobs.

The kidneys help regulate –

1. Blood pressure

2. The level of salts in the blood.

The kidneys clean the blood, removing waste products and extra water.

Kidneys produce those hormones that help regulate water and salt levels and maintain good blood pressure control.

It is the purifying system for blood as it **removes metabolic wastes and retains the proper amounts of water, salts, and nutrients.**

Due to Insulin Resistance, your pancreas doesn't make enough insulin or your body doesn't use insulin properly created by the pancreas. This leads to high blood sugar levels (**hyperglycemia**), which can damage the blood vessels in the kidneys and cause the filtering system (glomerulus) of the Kidneys to stop working effectively, which can lead to Chronic Kidney Failure.

Correction of Metabolic Acidosis improves insulin resistance in CKD.

Metabolic acidosis means there is too much acid in your blood and your kidneys cannot remove enough acid from the blood. People with CKD are at risk for metabolic acidosis, as their kidneys have more trouble removing acid from the blood.

Risk factors for metabolic acidosis for people with CKD are:

Hyperkalemia (high potassium levels)

Albuminuria/Proteinuria (too much albumin/protein in your urine)

Each Kidney is made up of millions of tiny filters called nephrons. When **hyperglycemia (High blood sugar) occurs**, it can cause Insulin Resistance, which can further damage the blood vessels and nephrons in the kidneys. It can also **develop high blood pressure**, which can damage kidneys.

Other than that, if the Liver releases too much **Urea (Ammonia), Uric acid, and Bilirubin (released from the destroyed red blood cells), then this Metabolic waste can lead to water Retention in the Kidney due to any of the above factors mentioned, which can lead to Kidney Failure or dialysis.**

Insulin Resistance causes Heart issues (CVD)

Your heart, a vital component of your circulatory system, is responsible for pumping blood throughout your body, regulating your heart rate, and maintaining blood pressure. Understanding its role is key to comprehending the circulatory system.

Blood, our body's lifeline, plays a crucial role in delivering oxygen, hormones, glucose, and other essential components to various parts of the body. The heart, in addition to its pumping function, ensures that adequate blood pressure is maintained in the body. The oxygen-carrying blood is referred to as oxygenated blood.

The circulatory system is complex, with two types of circulation: pulmonary and systemic. Understanding these types is key to understanding how blood flows through our body.

Pulmonary circulation carries deoxygenated blood away from the heart to the lungs and then it brings oxygenated blood back to the heart, whereas, in Systemic circulation, the oxygenated blood is pumped from the heart to every organ and tissue in the body, and deoxygenated blood returns to the heart.

A heart attack is when blood can't reach the heart.

Heart failure is when the heart fails to circulate blood properly.

Cardiac arrest is when the heart suddenly stops beating.

Congestive Heart Failure due to extra fluid (water retention) in different organs – The heart cannot pump enough blood.

Symptoms of a heart attack can be chest pain or tightness in the middle or heaviness, tightness across your chest, or pain in the left arm, but it may affect both arms and jaw.

What is ischemia?

Ischemia is a condition in which blood flow is restricted or reduced in a part of the body.

What is ischemic heart disease?

Narrowed heart arteries cause heart problems, so less blood and oxygen reach the heart muscle. This is also called coronary heart disease.

What is silent ischemia?

People can have ischemic issues without knowing it or having pain — silent ischemia (heart attack with no warning sign). People with angina also may have silent ischemia. People who have diabetes are also at high risk for **developing silent ischemia.**

Cardiovascular diseases (CVDs) are a group of disorders of the heart and blood vessels. They are as follows:

Coronary heart disease – a disease of the blood vessels supplying the heart muscle.

Congenital heart disease – **birth defects that disturb the normal development and functioning of the heart caused by deformities of the heart construction since birth.**

Deep vein thrombosis and pulmonary embolism are kind of **blood clots in the leg veins that can move to the heart or lungs.**

Endothelial Dysfunction

When Cholesterol/Fat/Sugar gets into the artery walls due to insulin resistance, it can make blood clot or raise blood sugar levels, and as a result, blood supply may stop due to blockage or create inflammation, which damages the lining of the inside arteries.

Insulin Resistance causes PCOS or PCOD

Polycystic ovary syndrome is a complex ovarian disorder, characterized by the formation of multiple cysts in the ovary.

The exact cause of polycystic ovary syndrome is not known, but the majority of women with polycystic ovary syndrome (PCOS) have insulin resistance or high insulin; their bodies can make insulin but can't use it effectively, increasing their risk for type 2 diabetes.

Insulin resistance, a condition where the body's cells don't respond normally to insulin, often causes high levels of insulin. This excess insulin leads the ovaries to produce too much testosterone, a hormone often associated with males. This hormonal imbalance interferes with the development of the follicles or prevents regular ovulation.

Too much insulin causes **hyperinsulinemia** and can also lead to **Type 2 diabetes**. Too much insulin is also an underlying problem of PCOS.

PCOS can disrupt or prevent ovulation, a process where eggs develop and are released from the ovaries. When this process is disrupted or prevented, it is called anovulation. Anovulation can lead to irregular or absent periods, one of the key symptoms of PCOS.

In a normal menstrual cycle, the Hypothalamus, a part of the brain, produces GnRH, a hormone that stimulates the Pituitary gland, a small gland at the base of the brain. The Pituitary gland then releases FSH (follicle-stimulating hormone) and LH (luteinizing hormone). These hormones activate the Ovaries for ovulation and stimulate the production of sex hormones—estrogen and Progesterone. This process is what leads to a regular menstrual cycle.

However, in a PCOS situation, the Hypothalamus does not produce GnRH, and this causes the FSH hormone level to decrease and the LH hormone to increase. An increase in LH hormone will elevate the Androgens Hormone (Male testosterone hormone in Ovary) levels and estrogen hormone levels and this will again decrease the FSH Hormone and this will trigger anovulation. The egg would not release and as a result, follicular cysts would be created. An anovulation will also decrease the Progesterone, and this will lead to a hike in LH Hormones. It is an endocrine imbalance which makes this continuous vicious cycle.

The symptoms of PCOS are:

- Missed periods, irregular periods, or very light periods.

- Amenorrhea (the absence of menstrual periods.

- Ovaries are large with many eggs' sac cysts.

- Excess body/facial hair (hirsutism).

- Weight gain around the waist area.

- Infertility.

- skin tags around the neck or armpits.

Insulin Resistance causes Thyroid (Disruption in Endocrine Hormones) - Autoimmune disease

T4 is the hormone produced in the thyroid gland, which is generally inactive. Low levels of T4 produce **hypothyroidism** (the most common form is Hashimoto thyroiditis, an auto-immune condition in which the body attacks its own thyroid cells). In **hypothyroidism**, the thyroid gland doesn't produce enough thyroid hormone, called an underactive thyroid. In **hyperthyroidism** (an overactive thyroid), the thyroid gland produces too much thyroid hormones.

T3 is a more biologically active hormone. Once your thyroid releases T4, certain organs in your body transform T4 (Thyroxine) into T3 (triiodothyronine) so that it can impact your cells and your metabolism.

Insulin resistance, a condition where your body's cells don't respond properly to insulin, can significantly impact your thyroid. It can disrupt thyroid hormone levels and function, potentially leading to health issues. The thyroid hormones T3 and T4, which are responsible for maintaining a fine balance of glucose in your body, are particularly affected.

Elevated insulin levels can interfere with the conversion of the inactive thyroid hormone (T4) to the active thyroid hormone (T3) in the body.

Hypothyroidism can break this equilibrium and alter glucose metabolism, which can lead to insulin resistance.

The menstrual cycle is regulated by the interaction of hormones: luteinizing hormone, follicle-stimulating hormone, and the female sex hormones estrogen and progesterone. If there is very little estrogen, then it can cause irregular and missed periods (amenorrhea), along with irregular thyroid hormones, which can cause a woman to miss her period.

Insulin Resistance causes Uterine fibroids

Insulin resistance can cause PCOS, and it can lead to high estrogen hormone, which will affect the increased risk of **Uterine fibroids and uterine Adenomyosis.**

If estrogen (female sex hormone) levels are too high. In that case, it can lead to fibroid development and growth, resulting in heavy and prolonged periods and bleeding between periods (Menorrhagia).

Estrogen Hormone influences puberty, the menstrual cycle, and menopause.

The underlying condition linked to High Estrogen (Hormonal Imbalance released by ovaries and the glands on the kidneys) levels are:

Insulin resistance

Polycystic ovary syndrome (PCOS) - Women who have PCOS symptoms, are often insulin resistant; their bodies can make insulin but can't use it effectively. PCOS is said to be caused by an imbalance in reproductive hormones (estrogen and progesterone, Follicle-stimulating hormone, luteinizing hormone).

Endometriosis

Uterine fibroids

Uterine Adenomyosis

Uterine adenomyosis is a chronic disorder causing heavy menstrual bleeding and infertility, and prolonged exposure to estrogen can be a contributing factor for Uterine adenomyosis.

Endometriosis is a condition where uterine-like tissue grows outside the uterus, ovaries, fallopian tubes or the intestines. It can cause severe pain due to irregular periods.

Insulin Resistance causes Infertility

Insulin resistance may result in **hyperinsulinemia**, which drives excessive ovarian androgen (excess testosterone) production and estrogen deficiency.

If Insulin is high in the body, it can impact sex hormones related to fertility.

Insulin resistance can cause delayed maturation of eggs due to **higher blood sugar** levels (Hyperglycemia) which can inhibit ovulation that can result in infertility.

Other than that, underlying conditions linked to High Estrogen (Hormonal Imbalance released by ovaries and the glands on the kidneys) levels are:

Insulin resistance

Polycystic ovary syndrome (PCOS) - Women who have PCOS symptoms, are often insulin resistant; their bodies can make insulin but can't use it effectively. PCOS is said to be caused by an imbalance in reproductive hormones (estrogen and progesterone, Follicle-stimulating hormone, luteinizing hormone).

Endometriosis

Uterine fibroids

Uterine Adenomyosis

Uterine adenomyosis is a chronic disorder causing heavy menstrual bleeding and infertility, and prolonged exposure to estrogen can be a contributing factor for Uterine adenomyosis.

Endometriosis is a condition where uterine-like tissue grows outside the uterus, ovaries, fallopian tubes or the intestines. It can cause severe pain due to irregular periods.

Insulin Resistance causes Male Sperm Count

Insulin Resistance, a condition where the pancreas fails to produce enough insulin or the body doesn't utilize it effectively, has a direct and detrimental effect on male fertility. This is due to the resulting high blood sugar levels (**hyperglycemia**), which can lead to blood vessel damage. In the context of diabetes, impaired sperm glucose metabolism can further damage nerves and blood vessels, **significantly reducing fertility.**

Men with pre-diabetes have low testosterone levels, resulting in damage to sperm DNA.

Dyslipidemia, a condition characterized by high LDL levels or triglycerides, along with insulin resistance, can significantly affect male fertility. This is due to the pathophysiology of sperm dysfunction associated with Metabolic syndrome, which is a known harmful factor in male fertility. It is often linked to Hypogonadism, decreased sperm concentration, and motility.

Male Hypogonadism is a condition when the sex glands/gonad (sex steroids in testis) produce little or no sex hormones (testosterone).

The testicles are also part of the endocrine system because they secrete testosterone hormones and production of sperm. If there are low sperm count (oligospermia), then it will cause male infertility.

Insulin Resistance causes Erectile Dysfunction or Low Libido (Low Sex drive)

The scrotum is a sac that holds the testicles. It has the arteries and veins that deliver blood to the reproductive glands but if there is swelling in the veins that drain the testicles, which is called varicocele.

Testosterone is an essential male hormone. In men, it is produced in the testicles. Testosterone is responsible for stimulating sperm production and regulating sex drive (libido).

In men, LH primarily stimulates testosterone production, while FSH stimulates the production of sperm.

Low testosterone may cause **Erectile Dysfunction** problem which is said to be associated with cardiovascular disease and is connected **with Endothelial Dysfunction** that results from **Insulin Resistance.**

Metabolic Syndrome (cardiovascular disease - **Endothelial Dysfunction**) leads to Insulin Resistance which results in **Sexual Dysfunction.**

Endothelium makes nitric oxide, which acts as a vasodilator and widens the blood vessels so that blood can flow freely. Insufficient nitric oxide can lead to narrowed blood vessels, which can contribute to high blood pressure and inflammation in the artery walls (**Atherosclerosis**—buildup of plaque in the artery walls due to fats, cholesterol, and other substances).

Therefore, Atherosclerosis in penile arteries leads to **Erectile Dysfunction** which is caused by **Insulin resistance.**

Insulin Resistance causes Diabetes Ketoacidosis (DKA) - Type 2 Diabetes

Those who have type 2 diabetes have a hard time using glucose (sugar) from food for energy due to Insulin Resistance which begins post we eat, and then carbohydrates in food break down into glucose which enters into the bloodstream, and blood sugar levels go up. And when it happens, the pancreas sends insulin into the blood.

Insulin helps open the cells throughout the body to let glucose in, giving them the energy they need.

However, if the cells don't respond to it (Insulin from the pancreas) as they should, then this condition is called **insulin resistance**, because the glucose can't get into cells, and as a result, the blood sugar level rises, then the pancreas works harder to make even more insulin, and eventually, over time, the pancreas can't keep up this continuous cycle and it gets exhausted, and the blood sugars stay high, and the person develops type 2 diabetes.

Therefore, if your cells become too resistant to insulin, it leads to elevated blood glucose levels (hyperglycemia) due to pancreatic beta cells, which make insulin and if it gets worn out then the pancreas no longer produces enough insulin to overcome the cells' resistance, which, over time, leads to pre-diabetes, higher blood glucose levels and Type 2 diabetes.

Insulin Resistance causes Fibrosis (scarring) of the lungs – Pulmonary Embolism

Insulin resistance, a condition where the body's cells become less responsive to insulin, and hyperinsulinemia, a state of high insulin levels in the blood, have been known to impair endothelial cell function, resulting in thrombosis, the formation of a blood clot. This can pose significant risk factors for cardiovascular and peripheral arterial disease by creating a blood clot in a deep vein, and it can cause serious complications, such as pulmonary embolism, where the clot travels to the lungs and blocks blood flow.

Increased blood clotting due to elevated fasting insulin levels, a common symptom of insulin resistance, may lead to pulmonary embolism, in which a blood clot blocks a vessel in the lungs. This highlights the direct impact of insulin resistance on cardiovascular health.

Pulmonary embolism is a condition where an embolus, usually formed in the leg (known as a deep vein thrombosis), travels from there to one of the lungs› arteries, causing a blockage.

Pulmonary edema (often caused by congestive heart failure when the heart cannot pump efficiently) is a condition in which fluid builds up in the lungs, making it difficult to breathe which can be caused by a pressure imbalance within the heart.

Insulin resistance, a metabolic syndrome, is directly linked to inflammation and oxidative stress, disruption of free radicals and antioxidants in your body that leads to cell damage. This connection leads to micro and macrovascular damage to the cardiovascular system, such as COPD, which is associated with a chronic systemic inflammatory condition that affects pulmonary blood vessels.

Insulin Resistance causes Hyperinsulinemia connection Insulin Injection

Hyperinsulinemia is connected to insulin resistance, a condition in which the cells in your liver and muscles don›t respond effectively to insulin. The pancreas makes too much insulin to overcome the resistance, leading to higher levels of insulin in the blood.

It's crucial to understand that if your pancreas can produce enough insulin to counteract your cells' poor response, your blood sugar levels will stay within a healthy range. This underscores the pivotal role of insulin in upholding your overall health.

However, if the cells in your body become too resistant to insulin, it can lead to **elevated blood glucose levels** (hyperglycemia). This, in turn, can cause prediabetes and eventually, Type 2 diabetes. It›s important to be aware of these potential consequences and take preventive measures.

Therefore, Hyperinsulinemia can cause abnormally high levels of insulin in your body. This can affect blood sugar levels and has strong links with type 2 diabetes.

In type -1 diabetes, the body does not make any insulin and therefore insulin has to be injected regularly.

In type-2 diabetes, insulin injection might be part of the treatment plan due to not having a healthy lifestyle and when other diabetes treatments don't control your blood sugar well enough.

Insulin is a vital treatment for type 2 diabetes. When the insulin your pancreas makes is not working properly, it produces more insulin and this can lead to the situation where pancreas worn out and unable to produce any more insulin. **In such cases, insulin injections are necessary to** regulate blood sugar levels (bring down blood sugar level).

Insulin Resistance causes Obesity

When hyperinsulinemia sets in, cells develop a resistance to insulin, leading to a surge in blood sugar levels. This, in turn, can pave the way for obesity and weight gain, as the excess glucose from the blood is not utilized for energy.

Fat (Lipid) that builds up deep in the abdomen raises the risk of **insulin resistance**, which could be due to **hyperinsulinemia, hyperglycemia, Inflammation, and dyslipidemia.**

One possible mechanism for the development of insulin resistance is the accumulation of visceral fat, which is fat that surrounds the organs in the abdomen. This type of fat is particularly harmful as it releases inflammatory cytokines, which can lead to lipid accumulation in the liver and muscle, further contributing to insulin resistance.

Cytokines are cell signaling proteins that control the growth and activity of other immune system cells and blood cells. Inflammatory cytokines trigger **Inflammation** and relay messages that coordinate the body›s immune response to fend off attackers, like germs.

The excessive accumulation of fat in adipose tissue can lead to the production of pro-inflammatory cytokines. These cytokines, which are cell signaling proteins, can inhibit the insulin signal, leading to insulin resistance in both adipose tissue and the **liver.**

Insulin Resistance causes Fits/Epilepsy/SNS activity (Cortisol Hormone due to Stress)

During stress, the hypothalamus sets off an alarm system that triggers the endocrine system to release adrenaline and cortisol hormones (from the adrenal glands found atop the kidneys), which will lead to the activation of the body's "fight or flight" hormones to protect against threats by keeping it on high alert.

In addition, cortisol (the stress hormone) triggers the release of sugar in the bloodstream from the liver to boost energy by reducing insulin release and increasing glycogen, and consequently blood glucose, during times of stress so that the brain can use the glucose (sugar).

SNS (sympathetic nervous system) is a physiological response to stress conditions (also known as the fight-or-flight response) that speeds up your heart rate and delivers more blood to areas of your body that need more oxygen. SNS stimulation in the pancreas causes a decrease in insulin secretion and an increase in blood glucose levels.

Increased sympathetic activity is also associated with developing insulin resistance and hypertension (metabolic abnormalities).

Acute sympathetic activation gives rise to insulin resistance and endothelial dysfunction. The long-term activation of the stress response system and too much exposure to cortisol can put you at higher risk of many health problems, which are:

- Headaches

- Epilepsy, Fits, Seizure

- Sleep problem

- Problems with memory and focus

- Anxiety, Depression

- Insomnia

- CVD

- Hypoxia (low blood supply or low oxygen content in the blood)

Insulin Resistance causes Auto Immune Disease (Rheumatoid Arthritis)

In Autoimmune disease, your immune system damages healthy cells in your body in an attempt to protect you from foreign cells (pathogens) that may cause disease and infections. It creates specific cells to target these foreign cells.

Autoimmune diseases may result from exposing the body to viruses or other environmental toxins. One of the common problems is Rheumatoid Arthritis, which is autoinflammatory arthritis.

Rheumatoid Arthritis is an immune system issue, and Gout is a potential autoimmune disorder caused by too much uric acid in the bloodstream.

Hyperuricemia (Elevated serum uric acid) is defined as an elevated serum uric acid level in the blood. It is a waste product created when your body breaks down chemicals called purines. The uric acid passes through your kidneys, but the body cannot excrete it properly. This can lead to Gout, a metabolic disease that results in acute intense inflammation of the joints of the legs or thumb/fingers.

Hyperuricemia (Elevated serum uric acid or Gout) can lead to a few metabolic syndromes, such as increased risk of hypertension, insulin resistance, diabetes, and cardiovascular disease.

Hyperuricemia with Hyperglycemia can cause insulin resistance because in Type 2 diabetes, if your body doesn't use insulin well and then sugar stays in the blood instead of moving into cells, then this condition is called insulin resistance and this can lead to the development of gout/ hyperuricemia which can make insulin resistance worse.

Insulin Resistance causes Diabetic Neuropathy

Diabetes is one of the biggest problems in the current time, which can cause diabetic retinopathy, diabetic peripheral Neuropathy, and diabetic nephropathy.

The most common symptoms of diabetic Neuropathy are burning feeling numbness and tingling which often begin in the feet or hands.

Diabetes can harm your nerves, which is called Neuropathy: Peripheral, Autonomic, Proximal, and Focal.

Peripheral Neuropathy affects the feet and legs, which leads to Tingling, Numbness and burning session.

Focal Neuropathy affects specific nerves, leading to Double vision, Eye pain, and Paralysis on one side of the face (Bell's palsy) and the lower back or leg.

People who have diabetes are more prone to risk for developing retinal diabetic Neuropathy, which leads to the loss of nerve cells in the retina.

Proximal Neuropathy causes pain in the thighs, hips and buttocks.

Autonomic Neuropathy affects the stomach, blood vessels, urinary system, and sex organs.

Insulin Resistance causes Pancreatic Problem

Insulin is a hormone made by your Pancreas that helps sugar enter your cells to be used as fuel. In people with insulin resistance, cells don't respond normally to insulin and glucose can't enter the cells as easily. As a result, your blood sugar levels rise, and the Pancreas releases more insulin to lower your blood sugar. Insulin resistance is the inability of cells to respond to insulin, which is the main feature of metabolic syndrome that leads to high blood sugar and type 2 diabetes.

In Insulin resistance, cells in your muscles, fat, and liver don't respond well to insulin (a hormone made by your Pancreas that helps manage your blood glucose). Once glucose (blood sugar) enters your bloodstream, insulin helps it get into your cells. However, if someone has insulin resistance, this process doesn't work well, and as a result, cells aren't letting glucose in, so more and more blood glucose piles up in your bloodstream.

The Pancreas keeps making insulin, and as a result, the cells become insulin-resistant, and blood glucose levels keep rising.

Insulin is a hormone that regulates the level of glucose in the blood and helps glucose enter the cells. If the body's response to insulin is impaired, cells cannot effectively utilize insulin to transport glucose from the bloodstream into the cells. However, when this natural process fails, the Pancreas produces more insulin to compensate for this resistance, leading to the problem of Insulin Resistance.

Due to Insulin Resistance, your Pancreas doesn't make enough insulin, or your body doesn't use insulin adequately created by Pancreas.

Insulin Resistance causes Prostate Enlargement

The prostate serves the purpose of nourishing semen and lubricating the urethra so sperm can travel outside of the body.

Hyperglycemia (High blood sugar) can damage blood vessels, impede blood flow, and can lead to the symptom **BPH (benign prostatic hyperplasia)** or prostate enlargement.

Hyperglycemia (caused by Insulin resistance) and BPH can worsen the symptoms of conditions such as a slower rate of urine flow.

Insulin resistance (Insulin-resistance syndrome is a group of disorders, such as diabetes, hyperinsulinemia, and sympathetic overactivity) predominantly increases the risks of **BPH and LUTS** (lower urinary tract symptoms). So, diabetic neuropathy and BPH (benign prostatic hyperplasia) can lead to dysfunctional bladders.

Another probable link between diabetes and BPH is the presence of insulin-like growth factor (IGF), which is produced by the liver and if there is any dysregulation of the insulin-like growth factor (IGF), that can also increase the risk of BPH and urinary tract infections (UTIs).

Insulin Resistance causes Cancer

Insulin resistance is a critical cause of metabolic dysfunctions. Metabolic dysfunction develops metastasis through Hyperglycemia, hyperinsulinemia, and insulin resistance, or chronic inflammation that can lead to cancer.

Hyperinsulinemia and Hyperglycemia result from insulin resistance, which happens when cells in the liver and muscles don't respond to insulin.

However, if you are insulin resistant, it leads to elevated blood glucose levels (Hyperglycemia), which may lead to diabetes. This can result in obesity, **inflammation**, and hormone imbalance, which raise the risk for multiple types of cancer, such as Liver Cancer, Pancreatic Cancer, Breast Cancer, and Prostate Cancer. So, Insulin resistance is one of the factors for causing cancer.

Pathogens like bacteria, viruses, fungi, chemicals, and radiation can cause inflammation, which can lead to cancer when the immune system fails to do its job and allows the cells to divide and grow.

The most common inflammatory diseases are:

Chronic inflammation can cause autoimmune diseases, like lupus, rheumatoid arthritis (RA), psoriasis and cardiovascular diseases,

Gastrointestinal disorders like inflammatory bowel disease, ulcerative colitis. Furthermore, Lung diseases, such as chronic obstructive pulmonary disease (COPD) and asthma.

Sometimes there is chronic inflammation in the body which can damage cell DNA and affect how cells grow and divide. That could lead to the growth of tumors and cancer. Most cancer deaths are caused by metastatic cancer, wherein cancer cells break away from the primary tumor (where they originated), travel through the blood or lymph system, and establish a new tumor in other organs or tissues of the body.

Insulin Resistance – Blood Cancer

Blood cancer is a type of cancer that affects your blood cells (white blood cells, red blood cells, platelets).

Leukemia, Lymphoma, and Myeloma are some of the most common types of blood cancer.

Blood cancer is caused by mutations in the DNA within blood cells, which cause them to start behaving abnormally. Leukemia is a type of blood cancer that affects white blood cells in the blood and bone marrow.

Leukemia affects blood cells, and it is the most common cancer in children. Children may experience anemia (a low number of red blood cells in the blood).

Acute lymphoblastic leukemia is a type of cancer of the blood and bone marrow that affects white blood cells to fight infection.

Acute lymphoblastic leukemia is also prevalent in small kids. It occurs when a bone marrow makes too many lymphocytes (a type of white blood cell).

Leukemia also leads to pulmonary leukemic infiltration, in which leukemic cells spread to other tissues within the lungs.

Insulin Resistance – Colorectal Cancer

Colorectal cancer occurs when normal cells that line the colon (also known as the large intestine, part of the digestive system or GI tract) or rectum (end of the colon) change uncontrollably and form a tumor. The tumor then metastasizes further into the wall of the colon or rectum and invades blood or lymph vessels.

Common types of tumors can erupt in the colon and rectum. These include:

- **Carcinoid tumors**

- **Gastrointestinal stromal tumors (GISTs)**

- **Lymphomas** are cancers of immune system cells. They mostly start in lymph nodes.

- **Sarcomas** can start in blood vessels or other connective tissues in the wall of the colon and rectum.

Colon cancer can spread to specific organs:

- Liver – It can result in fatigue, jaundice, swelling or abdominal bloating.

- Lungs –Metastatic colon cancer may spread to the lungs.

- Lymph nodes – Metastatic colon cancer can affect the lymph nodes in the abdominal area, resulting in abdominal bloating and swelling.

- Peritoneum – Metastatic colon cancer can result in abdominal pain.

Symptoms of colon cancer are rectal bleeding, blood in stool, abdominal pain, and fatigue.

Insulin Resistance causes Skin Disorder

Insulin resistance (Metabolic Syndrome) can induce skin disorder due to hyperinsulinemia, hyperglycemia, dyslipidemia, hypertension, and SNS, which affects the Immune System of the body.

Metabolic Syndrome adds stress to the body and makes nearly every system work harder, including the immune system's white blood cells.

This means that if you have Metabolic Syndrome/ autoimmune conditions, then your immune system releases excess aberrant cytokine (within adipose tissue) from adipocyte-associated results from insulin resistance, leading to SNS, hypertension, and inflammation.

Metabolic Syndrome **can lead to weak immune response by producing pro-inflammatory cytokines and inhibiting anti-inflammatory responses** which cause autoimmune disorder.

Skin diseases that have commonly been associated with insulin resistance and metabolic syndrome lead to following:

- Psoriasis

- Androgenetic alopecia

- Vitiligo

- Eczema

- Hives

- Melanoma (Skin Cancer)

- Lupus

Insulin Resistance causes IBS

IBS is a GI Tract disorder characterized by symptoms typically caused by Metabolic syndrome (Insulin Resistance) other than food and antibiotics.

Ulcerative colitis is an inflammatory bowel disease *(IBD) that* causes irritation, inflammation, and ulcers in the lining of the large Intestine or colon. *It is* caused by chronic inflammation in the digestive tract.

Metabolic syndrome and irritable bowel syndrome (IBS) have been linked to altered gut microbiota (**as both conditions have been found to be influenced by altered gut microbiota**), as hyperinsulinemia and hyperglycemia can damage (the nerves in your GI tract become damaged) the GI tract, which can impair the functionality of how food is digested and absorbed.

Over time, hyperinsulinemia, hyperglycemia, hypertension, and SNS can lead to complications in the gastrointestinal tract.

Insulin Resistance, a key component of metabolic syndrome, can directly contribute to the development of gastrointestinal symptoms:

- Excess carbohydrates/sugar in the diet may also cause intestinal bloating.

- Insulin Resistance makes it harder to properly break down sugars in the small intestine.

- Hormonal shifts play a key role in IBS.

Ref From-

Zhang, L., Yao, Z., & Ji, G. (2018). Herbal Extracts and Natural Products in Alleviating Non-alcoholic Fatty Liver Disease via Activating Autophagy. Frontiers in Pharmacology. https://doi.org/10.3389/fphar.2018.01459

Fatty Liver Treatment, Causes, Symptoms and Home Remedies | FoodrFitness. https://foodrfitness.com/fatty-liver-treatment

Diabetes and alcohol: MedlinePlus Medical Encyclopedia. https://medlineplus.gov/ency/patientinstructions/000968.htm

Lubawy, M., & Formanowicz, D. (2022). Insulin Resistance and Urolithiasis as a Challenge for a Dietitian. International Journal of Environmental Research and Public Health, 19(12), 7160.

Thyroid Hormone: What It Is & Function. https://my.clevelandclinic.org/health/articles/22391-thyroid-hormone

Chapter – 1

Mitochondrial Dysfunction

Mitochondria are the powerhouse of your body cells, producing the energy (called ATP) your body needs by processing the oxygen from the foods you eat.

Other than ATP production, the Mitochondria's function in the cell is to regulate intracellular Ca2 + (homeostasis—the process by which the body reacts to changes to keep the balance inside the body), biosynthesis of amino acids, nucleic acids, lipids, and purines, and the production of reactive oxygen species.

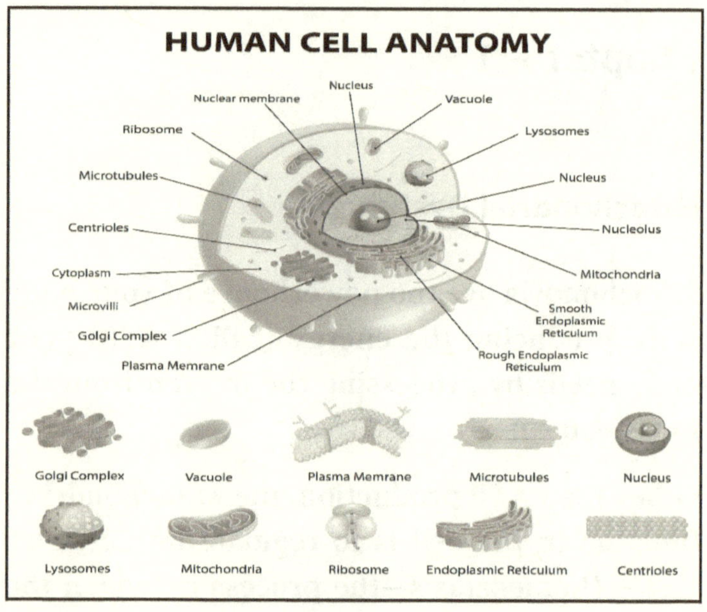

Photo Courtesy – Designed by Freepik

Mitochondria are present in every cell in the human body. Cells make up tissues and organs. The tissues or organs do not work correctly if the cells do not have enough ATP (adenosine-5′-triphosphate).

Mitochondrial Dysfunction is a condition in which the mitochondria in your cells cannot produce sufficient energy, which affects how the organs function in your body. Mitochondrial Dysfunction provides a clue for explaining cellular aging; it is a kind of organelle with a double membrane structure.

When a person has Mitochondrial Dysfunction, the mitochondria in the cells do not produce enough ATP energy. The severity of Mitochondrial Dysfunction depends on how many cells are affected and which tissue/organ in the body is impacted.

Any metabolic stress on the body causes Mitochondrial Disease. This may be because the cells are unable to cope with the extra demand.

Mitochondria are unique because they have their own DNA, called mitochondrial DNA, or mtDNA. Mutations in this mtDNA, or mutations in the nucleus of a cell, can cause mitochondrial disorder.

Different symptoms may occur depending on which cells within the body have disrupted mitochondria. Mitochondrial disease can cause a wide array of health concerns, such as fatigue, strokes, seizures, cardiomyopathy, cognitive function, diabetes mellitus, and liver, gastrointestinal, and kidney function.

If Mitochondrial Dysfunction affects the cells, it will affect tissues and then the related organs, which can result in excess fatigue and other symptoms that are common in almost every chronic disease.

Oxidative stress causes an overproduction of reactive oxygen species (ROS), which can lead to mitochondrial damage. The imbalance of ROS and antioxidants is disrupted by the overproduction of free radicals, which can damage DNA and degrade proteins and lipids.

Mitochondrial Dysfunction and oxidative stress are mainly involved in early aging signs: cancer, neurodegenerative disorders, and other metabolic syndromes (hypertension, hyperglycaemia, obesity, dyslipidaemia, hypertriglyceridemia, type 2 diabetes, cardiovascular disease, insulin resistance, etc.).

Metabolic syndrome, a cluster of conditions, is closely linked to obesity, high blood pressure, high triglycerides/LDL, low levels of HLD cholesterol, obesity/dyslipidemia, hypercoagulability, and insulin resistance.

It's important to understand that metabolic syndrome is not a disease itself but a group of risk factors that occur together, increasing the risk of CVD disease, stroke, and type 2 diabetes that can lead to Mitochondrial Dysfunction.

Mitochondrial Dysfunction symptoms include:

Neurodegenerative disease - Neurological problems such as seizures, Autism disorder, ADHD, Memory loss

Problem with Pancreas

Muscular dystrophy

Liver

Type 1 diabetes.

Multiple sclerosis

Heart

Thyroid and/or adrenal dysfunction

Gastrointestinal disorders

Kidney Failure

Ref From –

Mellitox Reviews 2024 • Consumer Report • Is it a Scam?. https://www.dumblittleman.com/mellitox-reviews/

Mitochondrial Disease | Children's Hospital of Philadelphia. https://www.chop.edu/conditions-diseases/mitochondrial-disease

Ji, J., Jiao, W., Chen, Y., Yan, H., Su, F., & L, C. (2018). ShenFu Preparation Protects AML12 Cells Against Palmitic Acid-Induced Injury Through Inhibition of Both JNK/Nox4 and JNK/NFκB Pathways. Cellular Physiology and Biochemistry. https://doi.org/10.1159/000487728

Emothion, Orally Absorbable Glutathione | Dr Craige Golding. https://drcgolding.co.za/emothion-orally-absorbable-glutathione/

Case of Metabolic Syndrome. https://brainmass.com/biology/human-anatomy-and-physiology/case-of-metabolic-syndrome-602282

What is Homeostasis

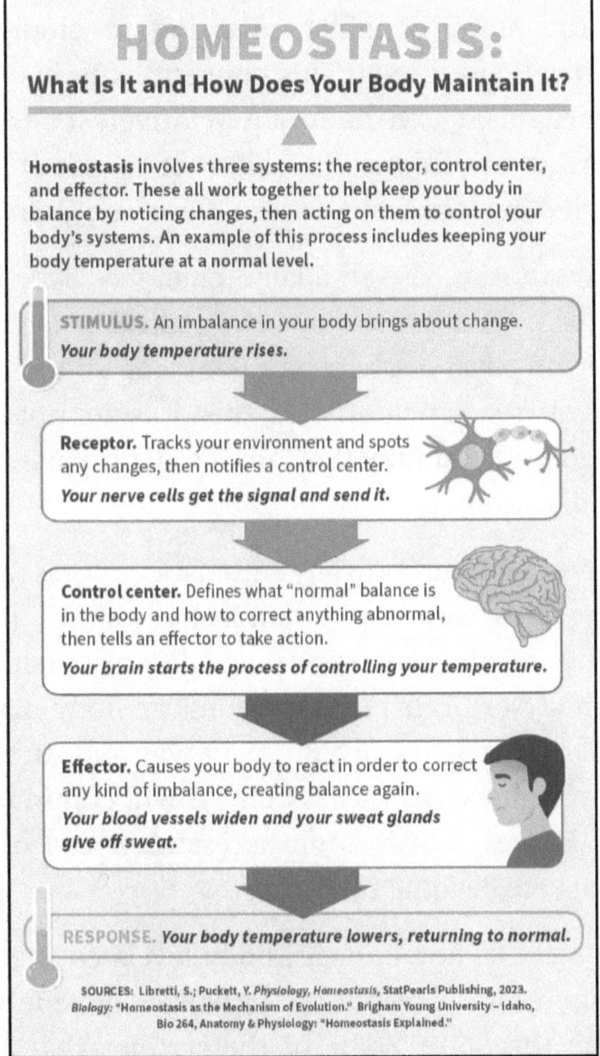

Picture Courtesy From - https://www.webmd.com/a-to-z-guides/what-is-homeostasis

Homeostasis is defined as a self-regulating process by which biological systems keep up the stability while adjusting to changing external conditions. This concept explains how an organism can maintain relatively constant internal conditions that allow it to adapt and survive in a changing external environment.

Homeostasis has gradually emerged over the centuries and has become the central organizing tenet of physiology. If one does not understand this self-regulating process, then it is not possible to comprehend fully the body's function in health and disease.

Homeostasis refers to any automatic process that a living thing uses to maintain its body steady from the inside while adjusting to conditions outside of the body or in its environment. The body makes these changes to work the right way and survive, and if it does this successfully, it will continue to live. However, if it is unsuccessful, it can cause imbalance, leading to disease.

In a state of homeostasis, body levels constantly rise and fall in response to changes outside and inside the body. Some of the systems that self-regulate to stay at normal levels are:

- Blood sugar

- Blood pressure

- Energy

- Acid levels

- Oxygen

- Proteins

- Temperature

- Hormones

- Electrolytes

Any bodily system in balance reaches a steady state that can withstand outside change. But if this system is disrupted, controlling devices built into your body react to create a new balance. One process is called feedback control. All methods that involve carrying out and controlling a function are examples of homeostasis. This occurs whether made possible by the nervous system, hormonal system, or electrical currents.

How does water help maintain homeostasis?

Water helps control body temperature by absorbing heat, spreading it among the body's liquid parts, and getting rid of it through sweat.

Water also plays a role in blood pressure management. For instance, the kidneys will retain water if blood pressure is too low.

Which body systems help maintain homeostasis?

Homeostasis involves all body organs, with different systems acting together to keep it in balance.

This includes:

- Integumentary system
- Nervous system
- Musculoskeletal system
- Cardiovascular system
- Endocrine system

What improves homeostasis?

Many things can help support homeostasis across different systems. For instance, plenty of sleep and nutritious food can improve your immune system's homeostasis. Exercise can help improve blood sugar homeostasis. You can also improve homeostasis, particularly in your immune and nervous systems, by avoiding ongoing stress through things such as coping skills, support, and meditation.

Difference between Mitochondrial DNA and Nucleolus DNA

Mitochondrial DNA is the circular chromosome found inside the cellular organelles called mitochondria. In the cytoplasm, mitochondria are the site of the cell's energy production and other metabolic functions. Offspring inherit mitochondria — and, as a result, mitochondrial DNA — from their mother.

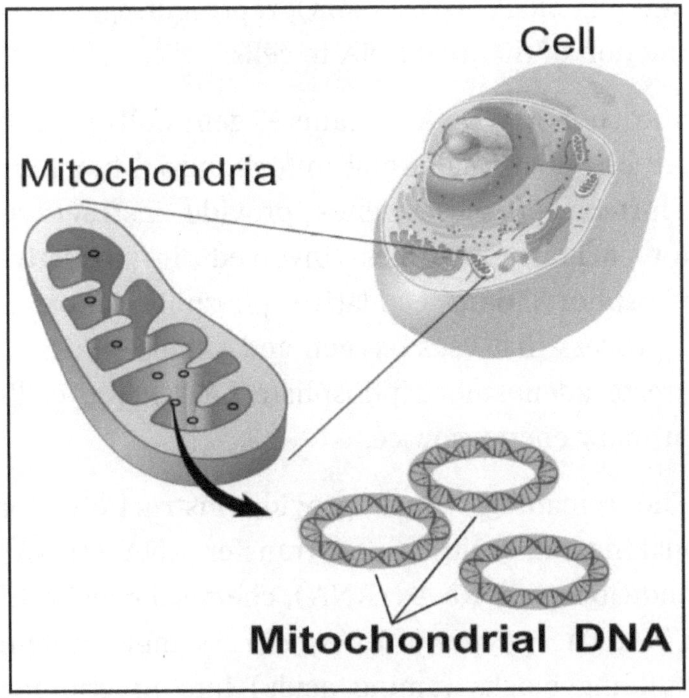

Picture Courtesy - https://en.wikipedia.org/wiki/Mitochondrial_DNA

Mitochondria are structures within cells that convert the energy from food into a form that cells can use. Each cell contains hundreds to thousands of mitochondria in the fluid surrounding the nucleus (the cytoplasm). Although most DNA is packaged in chromosomes within the nucleus, mitochondria also have a small amount of their own DNA. This genetic material is known as mitochondrial DNA or mtDNA. In humans, mitochondrial DNA spans about 16,500 DNA building blocks (base pairs), representing a small fraction of the total DNA in cells.

Mitochondrial DNA contains 37 genes, all of which are essential for normal mitochondrial function. Thirteen of these genes provide instructions for making enzymes involved in oxidative phosphorylation. Oxidative phosphorylation is a process that uses oxygen and simple sugars to create adenosine triphosphate (ATP), the cell's primary energy source.

The remaining genes provide instructions for making molecules called transfer RNA (tRNA) and ribosomal RNA (rRNA), chemical cousins of DNA. These types of RNA help assemble protein building blocks (amino acids) into functioning proteins.

As related to genomics, a nucleus is a membrane-enclosed organelle within a cell that contains the chromosomes. An array of holes, or pores, in the nuclear membrane allows for the selective passage of specific molecules (such as proteins and nucleic acids) into and out of the nucleus.

When you look at a picture of the cell, the nucleus is one of the most noticeable parts. It's in the middle of the cell, and the nucleus contains all of its chromosomes, which encode the genetic material. It is an integral part of the cell to protect.

The nucleus has a membrane around it that keeps all the chromosomes inside and distinguishes between the chromosomes inside the nucleus and the other organelles and components of the cell staying outside.

Sometimes, RNA needs to be trafficked between the nucleus and the cytoplasm, so pores in this nuclear membrane allow molecules to enter and exit the nucleus. It used to be thought that the nuclear membrane only allowed molecules to go out, but now it's realized that there is an active process also for bringing molecules into the nucleus.

Ref From –

https://www.webmd.com/a-to-z-guides/what-is-homeostasis

https://www.ncbi.nlm.nih.gov/pmc/articles/PMC7076167/

https://medlineplus.gov/genetics/chromosome/mitochondrial-dna/

https://www.genome.gov/genetics-glossary/Nucleus

Watanabe, C., Akhtar, W., Tou, Y., & Neittaanmäki, P. (2022). A new perspective of innovation toward a non-contact society - Amazon's initiative in pioneering growing seamless switching. https://doi.org/10.1016/j.techsoc.2022.101953

Frontiers | Homeostasis: The Underappreciated and Far Too Often Ignored Central Organizing Principle of Physiology.

https://www.frontiersin.org/articles/10.3389/fphys.2020.00200/full

https://www.webmd.com/a-to-z-guides/what-is-homeostasis

https://www.babymed.com/genetics/what-is-mitochondrial-dna

https://acelebrationofwomen.org/2022/04/world-dna-day-celebrated-april-25/

Nucleus. https://www.genome.gov/genetics-glossary/Nucleus

https://www.genome.gov/genetics-glossary/Mitochondrial-DNA#:~:text=Mitochondrial%20DNA%20is%20the%20circular,production%20and%20other%20metabolic%20functions.

Chapter – 2

Mitochondrial Dysfunction and Aging

Biological or cellular aging is related to a continuous decline in mitochondrial respiratory activity, i.e., mitochondrial dysfunction, which is responsible for reduced ATP production (responsible for tissue homeostasis and stress response) and elevation in oxidative stress that could damage proteins, lipids, and DNA.

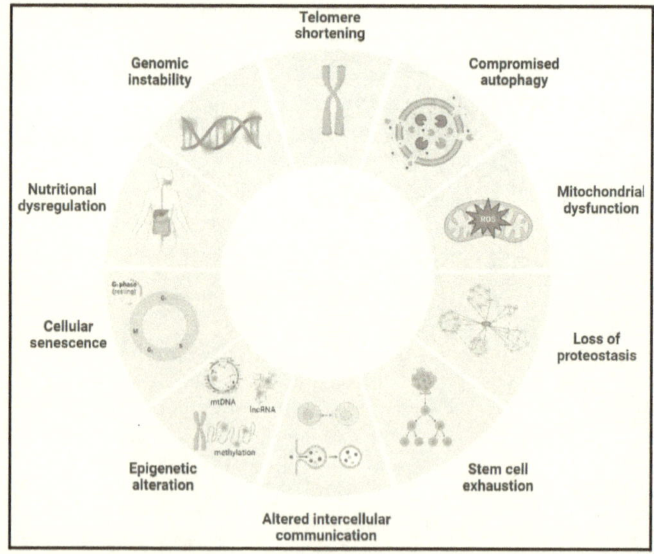

Photo Courtesy - (frontiersin.org) – (BioRender.com)

Mitochondrial dysfunction is one of the Hallmarks of Aging. As they age, mitochondria lose their ability to provide chemical ATP energy and release reactive oxygen species that harm Mitochondrial DNA (DNA Mutation) in cells.

One way to understand mitochondrial dysfunction is through inefficient ATP production, an essential fuel for cells of all types.

Aging is associated with the functional decline of diverse body organs, which can alter their capacity to cope with stress induced by metabolism, pathogens, and infections from an external toxic environment. This can damage the cellular macromolecules, posing a risk of several health issues that stimulate aging from cells to tissues to organs and give rise to cancer, cardiovascular disease, autoimmune disease, etc.

Aging is a degenerative process caused by accumulated oxidative stress that leads to cellular dysfunction by producing free radicals (natural byproducts of energy production) in mitochondria.

Mitochondrial ROS production increases with age because of a decline in mitochondrial function. When a mitochondrion is damaged, its ability to power the cell is impaired. This is caused by Free radicals, which are highly reactive and may damage proteins and DNA.

As this damage accumulates, the mitochondria become less effective at producing energy (ATP), leading to mitochondrial dysfunction.

Age-related changes in mitochondria are associated with a decline in mitochondrial function due to oxidative damage induced by reactive oxygen species (ROS). As aging progresses, NAD+ levels in human cells decrease. Increased ROS, DNA mutations, and oxidized proteins cause mitochondrial dysfunction in aging.

Mitochondrial dysfunction with increased oxidative damage contributes to biological aging. Antioxidants on Cellular levels and free radical scavengers that protect against this damage also decline with age. Subsequently, mitochondria within our cells become dysfunctional over the years, and they do not produce ATP as efficiently as they should. The accumulation of dysfunctional mitochondria also depletes an essential molecule called nicotinamide adenine dinucleotide (NAD).

Ref From –

https://www.frontiersin.org/journals/
endocrinology/articles/10.3389/
fendo.2024.1361289/full#f1

Rangarajan, S., Kurundkar, D., Kurundkar, A., Locy, M., Bernard, K., Deshane, J., & Thannickal, V. (2016). A70 THE GOOD, THE BAD, AND THE UGLY OF AGING: Mitochondrial Uncoupling Protein Ucp2 Regulates Aging-Related Experimental Lung Fibrosis. American Journal of Respiratory and Critical Care Medicine, 193(), 1.

Cantemir-Stone, C., Wang, Y., Evans, R., Bolon, B., LaPerle, K., & Marsh, C. (2014). A29 NEW KIDS ON THE FIBROTIC BLOCK: LUNG FIBROSIS AND FIBROBLAST BIOLOGY: Identifying Molecular Regulators Of Aging In Lung Fibrosis. American Journal of Respiratory and Critical Care Medicine, 189(), 1.

Cormio, A., Musicco, C., Pesce, V., Villani, R., Antonelli, A., & Barret, E. (2018). Age, mitochondria and bladder cancer. https://core. ac.uk/download/196286807.pdf

https://www.gowinglife.com/what-is-mitochondrial-dysfunction-the-hallmarks-of-ageing-series/

https://www.lifespan.io/topic/mitochondrial-dysfunction/

Chapter – 3

Anabolic vs Catabolic

Anabolism and catabolism are the two broad biochemical reactions that comprise metabolism. Metabolism shall consist of two significant parts; one is anabolism, and the other is catabolism.

Mitochondria are essential cell organelles that break down complex molecules into simpler molecules to release chemical ATP energy through the Catabolic Process.

The Anabolic Process consumes this energy to build complex molecules from simpler organic compounds (proteins from amino acids; carbohydrates from sugars or starch or fiber; fats (Triglycerides) from fatty acids and glycerol).

The term metabolism indicates the sum of the chemical reactions necessary to keep an organism alive. Cellular metabolism is composed of two distinct pathways: catabolism and anabolism.

Catabolic reactions break down macromolecules (i.e., sugars, lipids, and proteins) in their building blocks (glucose, fatty acids, amino acids) to release the energy in their chemical bonds and store it in the form of ATP. ATP is the acronym for adenosine triphosphate. It is an adenosine molecule linked to a chain of three phosphate groups (alpha, beta, or gamma, from the closest to furthest from the ribose sugar).

"Anabolic" and "Catabolic" sound similar but are opposites. Remembering the difference may help one think about how "anabolic steroids" promote the buildup of muscle mass. All of the complex molecules of life — carbohydrates, lipids, proteins, nucleic acids — are generated by anabolic reactions. Anabolic reactions are central to photosynthesis, protein synthesis, and DNA replication.

Mitochondria are the sites of cellular respiration (aerobic catabolic reactions), and they are, thus, essential in cellular bioenergetics. They are the primary producers of ATP molecules and the site where the intermediate metabolites of the catabolic reactions can be reintegrated into anabolic pathways to synthesize new macromolecules. Cellular respiration consists of three related series of responses.

First, sugars, peptides, and fatty acids are processed to generate acetyl-CoA, a key node in metabolism. It represents the main entry point to the subsequent mitochondrial energy conversion processes.

- Glucose is the monosaccharide most typically used for energy production. It undergoes a series of ten anaerobic lytic reactions in the cytoplasm, known as glycolysis. This will generate two net ATP molecules, two molecules of NADH (an essential source of reducing equivalents for redox reactions), and two pyruvate molecules. Pyruvate is then transported into mitochondria and converted into acetyl-CoA.

- Fatty acids are transported into the mitochondria via the carnitine-based system and then catabolized by β-oxidation to eventually generate acetyl-CoA and FADH2 and NADH cofactors

- The catabolism of some amino acids (e.g., phenylalanine, tyrosine, leucine, lysine, and tryptophan) can lead to acetyl-CoA or other intermediate molecules of the TCA cycle.

The actual "cellular respiration" starts with the **TCA cycle** (tricarboxylic acid cycle), also known as the **citric acid cycle** or, after its discoverer, the **Krebs cycle**. As all these denominations suggest, it consists of a circular series of biochemical reactions, all taking place in the mitochondria, resulting in the generation of one molecule of GTP (equivalent to ATP in terms of energy charge) and, most importantly, three molecules of NADH, and one molecule of FADH2, later used for the oxidative phosphorylation process. The TCA cycle initiates with the reaction of acetyl-CoA and oxaloacetate to form citrate.

Ref From –

https://www.enzo.com/note/what-is-the-role-of-mitochondria-in-cellular-metabolism-and-bioenergetics/#:~:text=Anabolic%20 reactions%20generally%20correspond%20 to,structures%2C%20tissues%2C%20and%20 organs.

https://biology.stackexchange.com/ questions/67442/catabolic-and-anabolic-reactions

Chapter – 4

Telomere and Mitochondrial Dysfunction

Mitochondrial and telomere functions have mostly been studied independently. In recent years, however, it has become clear that there are intimate links between mitochondria, telomeres, and telomerase subunits. Mitochondrial dysfunctions cause telomere attrition, while telomere damage leads to reprogramming of mitochondrial biosynthesis and mitochondrial dysfunctions, which have important implications in aging and diseases.

The aging field has expanded dramatically in the last three decades since the discovery that it is controlled, at least to some extent, by evolutionarily conserved pathways (Kenyon et al., 1993). Nine hallmarks of aging have since been postulated.

The primary causal hallmarks are genomic instability, telomere damage, and epigenetic alterations, which are partially overlapped or intertwined (Lopez-Otin et al., 2013). Telomere damage, for example, also leads to genomic instability (Fumagalli et al., 2012; Hewitt et al., 2012).

Another hallmark is mitochondrial dysfunction, which is also closely related to genomic instability, bacause cells have two genomes: the nuclear and mitochondrial.

In addition, mitochondria are where most intracellular reactive oxygen species (ROS) are produced (Wallace, 2005). The free radical theory of aging suggests that ROS causes oxidation damage in both mitochondrial DNA (mtDNA) and nuclear DNA, leading to the accumulation of mutations and eventually aging (Harman, 1956). It is well known now that the theory is oversimplified and is only part of the story, as ROS may also activate compensatory pathways that may negate their deleterious effects (Yee et al., 2014; Wang and Hekimi, 2015).

Telomere shortening is a hallmark of aging and stress-induced senescence. Telomere length critically affects cell senescence and organism life span. Accordingly, prevention of age-associated telomere shortening increases health and longevity.

A second hypothesis connecting hair graying with ischemic heart disease involves telomere shortening. Telomere length decreases with age, but telomere shortening has associated with oxidative stress and cardiovascular risk factors such as smoking and adiposity (Bojesen, 2013). Recently, telomere shortening associated with an increased risk of coronary heart disease and early death (Haycock et al., 2014; Weischer et al., 2012).

Senescence Cell

Role of cellular senescence in type 2 diabetes mellitus

The below figure depicts factors known to lead to senescence in cells: mitochondrial dysfunction, endoplasmic reticulum (ER) stress, and increased production of reactive oxygen species (ROS). Additional factors include telomere shortening, DNA damage, hyperglycemia, insulin resistance, and inflammatory signals. The picture reflects some of the reported adverse effects of cellular senescence: secretion of the senescence-associated secretory phenotype (SASP), which can promote the entry to neighboring cells' senescence and dysfunction. The proliferation of cells is suppressed, and they downregulate hallmark identity genes. Due to these characteristics, senescent cells can promote diabetes and metabolic diseases in multiple ways.

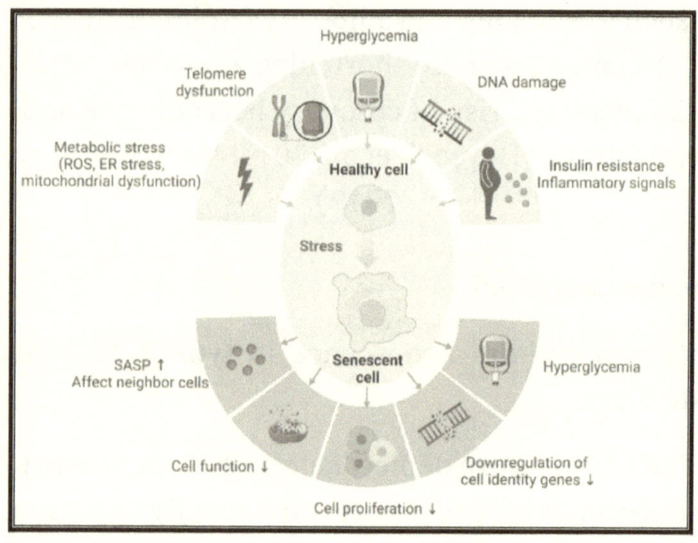

Pic Courtesy from - https://www.e-dmj.org/journal/view.php?number=2733
Pic Created in BioRender

Hyperglycaemia accelerates cellular senescence through multiple pathways. Therefore, senescence is an important cellular mechanism to consider in the pathophysiology of type 2 diabetes mellitus (T2DM). When normal cells are subjected to severe DNA damage, they either die by apoptosis or undergo irreversible cell proliferation arrest by induction of cellular senescence.

These biological defence mechanisms prevent the proliferation of abnormal cells that have suffered DNA damage. *Cellular senescence* is when cells irreversibly stop proliferating while retaining their metabolic activity. It can be induced by external stressors such as aging, obesity, and radiation due to DNA damage, telomere shortening, and mitochondrial dysfunction.

Cellular senescence is a process that results from a variety of stresses and leads to a state of irreversible growth arrest. Senescent cells accumulate during aging and have been implicated in promoting a variety of age-related diseases.

Cellular senescence may play an essential role in tumour suppression, wound healing, and protection against tissue fibrosis; however, accumulating evidence that senescent cells may have harmful effects in vivo and may contribute to tissue remodelling, organismal aging, and many age-related diseases also exists.

Various intrinsic and extrinsic factors can induce cellular senescence. Cellular senescence has beneficial biological functions in regulating embryonic development, wound healing, resolution of fibrosis, and tumour suppression. However, prolonged senescence can result in deleterious sequelae, including tumor development, chronic inflammation, immune deficit, and stem cell exhaustion. **Cellular senescence (CS) in the cardiovascular system** is driven by biomolecular alterations, primarily mitochondrial dysfunction, causing oxidative and glycative stress, resulting in telomere shortening, DNA damage, and telomere damage

Influencing factors:

- Chronological age, lifestyle habits, genetic inheritance, epigenetic changes, and psychosocial influences.

Types of senescence:

- In cardiovascular diseases, macromolecular damage leads to telomere length-dependent (replicative senescence) and length-independent (stress-induced premature senescence) senescence.

Cellular effects:

- Mitotic cardiovascular cells, such as endothelial and smooth muscle cells, undergo progressive telomere loss-based senescence, which is associated with cardiovascular disease phenotypes like atherosclerosis.

Potential implications:

- Atherosclerosis, heart rupture post-AMI (acute myocardial infarction), cardiac aging, abdominal aortic aneurysm, hypertension, heart regeneration, and cardiac remodelling.

Cellular senescence is a significant contributor to impaired bone strength, leading to conditions such as osteoporosis, fragility fractures, and sarcopenia.

Senescent cells in bone:

- Senescent osteoblasts exhibit diminished mineralization capacity and compromised functionality.

- Senescent osteoclasts become hyperactive, leading to bone loss and increased susceptibility to fragility fractures.

- Senescence-related modifications in bone marrow stem cells contribute to reduced osteogenic potential and impaired fracture healing.

Senescent alterations in muscle cells and satellite cells:

- Senescent muscle cells exhibit impaired stem cell self-renewal, alterations in the intracellular microenvironment, and increased protein degradation, contributing to sarcopenia.

- Senescence affects satellite cells, hindering their ability to sustain and rejuvenate through persistent p38 MAPK activity and p16INK4a overexpression.

Potential implications:

- A standardized set of senescence-associated factors and genes, called SenMayo, has been formulated to guide senotherapeutic therapies.

Cellular senescence plays a pivotal role in age-related disorders, including **Alzheimer's disease (AD) and Parkinson's disease (PD)**. CS is a potential contributor to age-associated inflammation in the brain, primarily in glial cells capable of replication.

AD pathophysiology:

- Senescent cells in AD contribute to neuroinflammation by releasing pro-inflammatory cytokines, leading to microglial activation and chronic inflammation.

- Senescent cells may impair clearance mechanisms, contributing to the accumulation of toxic protein aggregates.

PD pathophysiology:

- PD is characterized by dopaminergic neuron degeneration, abnormal protein accumulation, neuroinflammation, mitochondrial dysfunction, oxidative stress, and cellular senescence.

- Senescent cells in PD secrete pro-inflammatory molecules, contributing to chronic neuroinflammation and the loss of dopaminergic neurons.

Impact on brain structure and function:

- Senescent glial cells and neurons lead to structural and functional alterations in the brain. Disruption of cell-to-cell contacts affects neuron-glia interactions, impacting neuronal ion levels and metabolic homeostasis.

Therapeutic considerations:

- Interventions targeting cellular senescence present a promising avenue to prevent cell loss and preserve brain functionality.

Ref From –

https://www.ncbi.nlm.nih.gov/pmc/articles/PMC9362342/#:~:text=Cellular%20senescence%20has%20beneficial%20biological,deficit%20and%20stem%20cell%20exhaustion.

https://onlinelibrary.wiley.com/doi/full/10.1111/joim.13775#:~:text=Primarly%2C%20senescent%20induced%20replication%20arrest,of%20typical%20cellular%20protein%20biomarkers.

https://www.hmpgloballearningnetwork.com/site/wounds/article/cellular-senescence-what-why-and-how#:~:text=Cellular%20senescence%20is%20a%20process,variety%20of%20age%2Drelated%20diseases.

https://www.e-dmj.org/journal/view.php?number=2733

https://www.e-dmj.org/journal/view.php?number=2733

https://www.fightaging.org/archives/2023/03/cellular-senescence-in-type-2-diabetes/

https://thehighwire.com/editorial/revealed-vaccine-induced-cellular-aging/

Tikhonova, M., Chang, H. M., & Singh, S. (2023). Editorial: Experimental and innovative approaches to multi-target treatment of Parkinson's and Alzheimer's diseases - Volume II. Frontiers in Neuroscience, (), .

https://www.frontiersin.org/journals/cell-and-developmental-biology/articles/10.3389/fcell.2019.00274/full

https://www.sciencedirect.com/topics/biochemistry-genetics-and-molecular-biology/telomere-shortening

Telomeres Shortening with Age

Telomeres, the structures at the end of chromosomes, help to protect DNA from damage and allow chromosomes to replicate correctly during cell division.

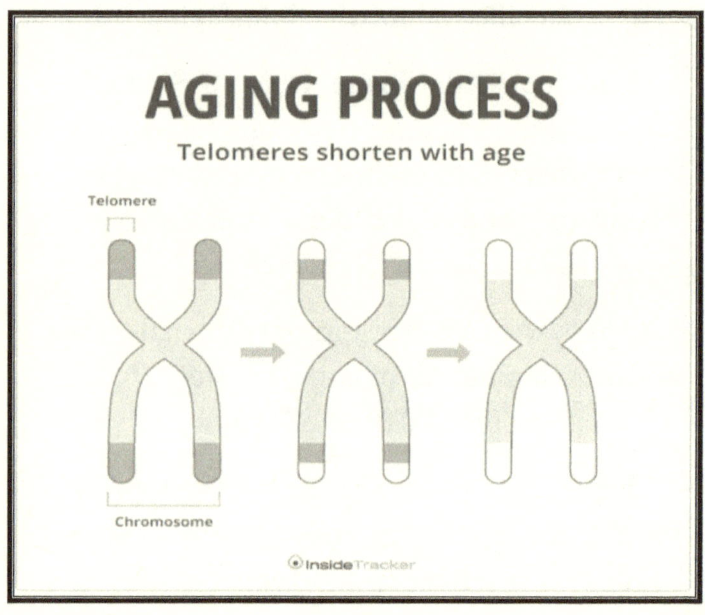

Telomeres, the specific DNA–protein structures found at both ends of each chromosome, protect the genome from nucleolytic degradation, unnecessary recombination, repair, and intrachromosomal fusion. Telomeres, therefore, play a vital role in preserving the information in our genome. A small portion of telomeric DNA is lost with each cell division as a normal cellular process. When telomere length reaches a critical limit, the cell undergoes senescence and/or apoptosis. Telomere length may, therefore, serve as a biological clock to determine the lifespan of a cell and an organism. Certain agents associated with specific lifestyles may expedite telomere shortening by inducing damage to DNA in general or, more specifically, at telomeres and may, therefore, affect the health and lifespan of an individual.

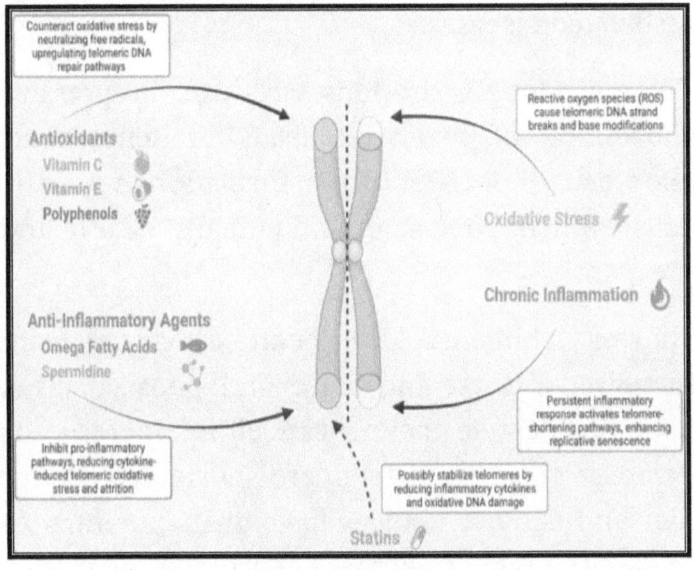

Pic Courtesy - https://www.frontiersin.org/files/Articles/1339317/fragi-05-1339317-HTML/image_m/fragi-05-1339317-g001.jpg

There has been growing evidence that lifestyle factors may affect an individual's health and lifespan by affecting telomere length. This review aimed to highlight the importance of telomeres in human health and aging and summarize possible lifestyle factors that may affect health and longevity by altering the rate of telomere shortening.

Recent studies indicate that telomere length, which can be affected by various lifestyle factors, can affect the pace of aging and the onset of age-associated diseases.

Telomere length shortens with age. Progressive shortening of telomeres leads to senescence, apoptosis, or oncogenic transformation of somatic cells, which affects an individual's health and lifespan.

Shorter telomeres have been associated with increased disease incidence and poor survival. Specific lifestyle factors can either increase or decrease the rate of telomere shortening. Better diet and activity choices have great potential to reduce the rate of telomere shortening or at least prevent excessive telomere attrition, leading to delayed onset of age-associated diseases and increased lifespan.

Certain lifestyle factors such as smoking, obesity, lack of exercise, and consumption of unhealthy diet can increase the pace of telomere shortening, leading to illness and/or premature death. Accelerated telomere shortening is associated with the early onset of many age-associated health problems, including coronary heart disease, heart failure, diabetes, increased cancer risk, and osteoporosis.

Individuals whose leukocyte telomeres are shorter than the corresponding average telomere length have a three-fold higher risk of developing myocardial infarction. Evaluating telomere length in elders shows that individuals with shorter telomeres have a much higher mortality rate than those with longer telomeres.

Smoking and obesity seem to have adverse effects on telomeres and aging. Smoking may expedite telomere shortening and the process of aging. Excessive telomere shortening can also lead to genomic instability and tumorigenesis. Consistently, most cancer cells' telomeres are shorter than normal cells. Smoking is associated with accelerated telomere shortening.

Obesity is associated with excessive telomere shortening, increased oxidative stress, and DNA damage. Furukawa *et al.* showed that waist circumference and BMI significantly correlate with elevated plasma and urinary levels of reactive oxygen species. Song *et al.* have shown that BMI strongly correlates with biomarkers of DNA damage, independent of age. The obesity-related increased oxidative stress is probably due to a deregulated production of adipocytokines.

Environment, nature of profession, and stress can also affect the rate of telomere shortening and health. Exposure to harmful agents and the nature of the profession may affect telomere shortening.

Stress is associated with the release of glucocorticoid hormones by the adrenal gland. These hormones have been shown to reduce the levels of antioxidant proteins and may, therefore, cause increased oxidative damage to DNA and accelerated telomere shortening.

What we eat and how much we eat can significantly affect our telomeres, health, and longevity.

A study by Farzaneh-Far *et al.* indicates that a diet containing antioxidant omega-3 fatty acids is associated with a reduced rate of telomere shortening. In contrast, lacking these antioxidants correlates with an increased telomere attrition rate in study participants.

The authors followed omega-3 fatty acid levels in blood and telomere length in these individuals over five years. They found an inverse correlation, indicating that antioxidants reduce the rate of telomere shortening.

Exercise can reduce harmful fat and help mobilize waste products for faster elimination, reducing oxidative stress and preserving DNA and telomeres.

Telomeres shorten with age, and progressive telomere shortening leads to senescence and/ or apoptosis. Shorter telomeres have also been implicated in genomic instability and oncogenesis. Older people with shorter telomeres have a three- and eight-times increased risk of dying from heart and infectious diseases, respectively. The rate of telomere shortening is, therefore, critical to an individual's health and pace of aging. Smoking, exposure to pollution, a lack of physical activity, obesity, stress, and an unhealthy diet increase the oxidative burden and the rate of telomere shortening.

To preserve telomeres and reduce cancer risk and pace of aging, we may consider eating less, including antioxidants, fiber, soy protein, and healthy fats (derived from avocados, fish, and nuts) in our diet, and staying lean, active, healthy, and stress-free through regular exercise and meditation. Foods such as tuna, salmon, herring, mackerel, halibut, anchovies, cat-fish, grouper, flounder, flax seeds, chia seeds, sesame seeds, kiwi, black raspberries, lingonberry, green tea, broccoli, sprouts, red grapes, tomatoes, olive fruit, and other vitamin C-rich and E-rich foods are a good source of antioxidants.

Key points

- Telomere length shortens with age.

- The rate of telomere shortening may indicate the pace of aging.

- Lifestyle factors such as smoking, lack of physical activity, obesity, stress, exposure to pollution, etc., can potentially increase the rate of telomere shortening, cancer risk, and pace of aging.

- Dietary restriction, appropriate diet (high fiber, plenty of antioxidants, lean/low protein, adding soy protein to diet), and regular exercise can potentially reduce the rate of telomere shortening, disease risk, and pace of aging

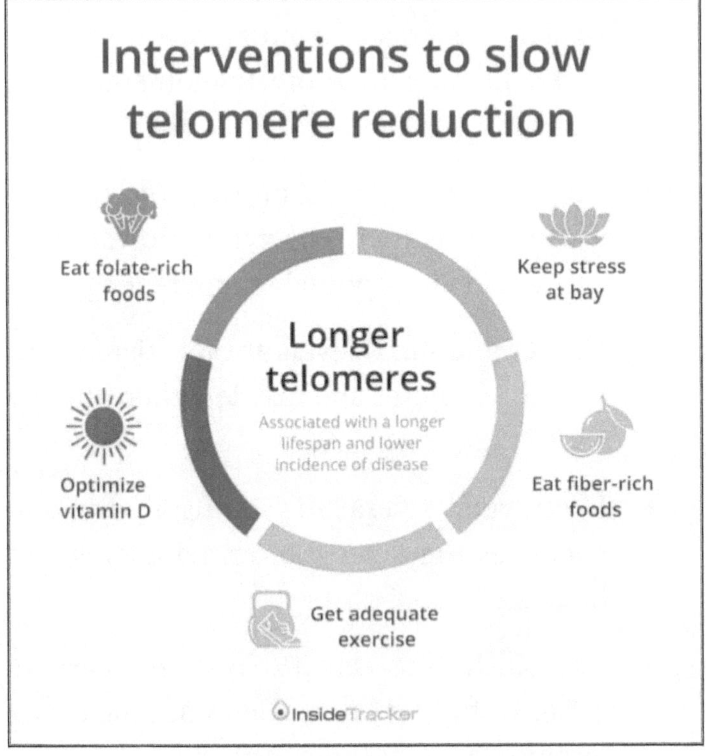

- Telomeres are the structures at the end of chromosomes, and mounting evidence suggests that telomere length declines with age—a change that scientists now consider a hallmark of aging.

- Focusing on a plant-rich diet full of fiber, antioxidants, vitamins, and phytonutrients may result in longer telomeres.

- Physical activity and exercise decrease oxidative stress and inflammation in the body, helping to protect telomeres from damage.

- Choose naturally folate-rich foods over those fortified with its synthetic variation (folic acid) for optimal telomere length.

- Higher vitamin D levels are associated with longer telomeres and may lengthen already shortened telomeres.

- Keep your stress and cortisol levels in check, as they can shorten telomeres and lifespan.

- Use InsideTracker to test biomarkers related to telomere length, including vitamin D, folate, and cortisol.

As telomere length shortens, so does your life expectancy. A recently published study found that having shorter telomeres increases your heart disease mortality risk by 3.18-fold and your infectious disease mortality risk by 8.54-fold. Researchers have discovered several methods for preserving telomere length and even growing it.

Antioxidants

Antioxidants neutralize free radicals, which can damage cellular structures (Halliwell, 1996), including DNA and telomeres. This protective effect can be essential in countering telomere shortening, possibly delaying cellular aging (Fenech et al., 2023).

B Vitamins Preserve Telomere Length

B vitamins, including vitamins B6, B12, and folate, are essential factors in the metabolism of the molecules that make up DNA. Their importance is, therefore, critical in supporting normal cellular replication. Low levels of B vitamins are common in aging adults and are closely associated with the risk of developing age-related diseases.

Homocysteine is a molecule associated with increased cardiovascular disease risk and poor blood vessel function. B vitamins are necessary for regular homocysteine disposal, which accumulates under conditions of B vitamin deficiency. Both elevated homocysteine levels and diminished B vitamin levels are closely associated with premature shortening of telomeres, leading to accelerated cellular aging. Thus, homocysteine-induced telomere shortening may be the central connection between B vitamin deficiency, high homocysteine levels, and cardiovascular disease.

It has long been known that B vitamin supplementation reduces homocysteine levels, and it was recently shown that people with low B vitamin levels have shorter telomeres.

Taken together, these findings suggest that keeping B vitamins at adequate levels effectively lowers toxic homocysteine levels and supports longer telomeres.

Vitamin D Promotes Telomerase Activity

Vitamin D has become known as one of the most versatile of nutrients. Vitamin D receptors are found on cells throughout the body, suggesting that more functions await discovery.

Recently, a molecular link was found between vitamin D and DNA repair, an action required for maintaining telomere length. Higher plasma vitamin D levels have been associated with longer telomeres. These findings have triggered other studies of vitamin D and its role in telomere function.

Vitamins C and E Preserve Telomere Length

Vitamin C, a potent water-soluble antioxidant, scavenges free radicals in the aqueous cellular environment, preventing damage to critical biomolecules. Notably, it can also augment the enzymatic action of telomerase, potentially supporting telomere elongation (Furumoto et al., 1998).

Vitamin E is a lipid-soluble antioxidant, primarily located in cell membranes, with its primary role to protect polyunsaturated fatty acids (PUFAs) from lipid peroxidation, which is a significant source of DNA damage, including telomeres (Corina et al., 2019). In the CORDIOPREV study, involving 1.002 cardiovascular disease patients, dietary vitamin E intake was found to significantly influence leukocyte telomere length, a biomarker for cellular aging. Patients with inadequate vitamin E intake exhibited shorter telomere length than those with sufficient intake (Corina et al., 2019).

Polyphenols

Polyphenols, widely present in sources such as fruits, vegetables, or tea, exert antioxidant effects by neutralizing oxidizing species. Resveratrol, a polyphenol found highly concentrated in berries, nuts, grapes, and red vine, exerts antioxidant and anti-inflammatory effects (Rotondo et al., 1998; Queiroz et al., 2009; Meng et al., 2021).

Anti-inflammatory agents

Prolonged inflammatory responses can increase oxidative stress and DNA damage, potentially accelerating telomere attrition. Several studies have investigated anti-inflammatory agents and their influence on telomere biology, being discussed in the following section.

Omega-3 fatty acids

Omega-3 fatty acids, particularly eicosapentaenoic acid (EPA) and docosahexaenoic acid (DHA), are essential polyunsaturated fatty acids (PUFAs) recognized for their anti-inflammatory properties. Emerging research indicates a potential role for these fatty acids in telomere biology (Chang et al., 2020; Ogłuszka et al., 2022).

Ref From –

https://www.ncbi.nlm.nih.gov/pmc/articles/PMC3370421/#:~:text=Dietary%20restriction%2C%20appropriate%20diet%20(high,risk%2C%20and%20pace%20of%20aging.

https://avc.com/2017/01/video-of-the-week-what-is-dna-and-how-does-it-work/

https://papyrus.bib.umontreal.ca/xmlui/bitstream/1866/25212/2/Paredes_Bonilla_Rosalba_Vivian_2020_memoire.pdf

https://doi.org/10.5772/38254

https://doi.org/10.1097/mco.0b013e32834121b1

https://youholistic.com/peptide-therapy/

https://iamzuri.com/2020/07/the-telomere-tell-all-diving-into-the-fountain-of-youth/

https://www.thegeorgetowndish.com/articles/nuts-and-seeds-reduce-aging-increasing-telomere-length

https://foodfactsandfads.com/tag/aging/

https://glennbolton.com/category/health/ageing-health/

https://ellendolgen.com/2019/12/news_from_north_america_menopause_society/

https://blog.insidetracker.com/strategies-slow-telomere-reduction

https://www.insidetracker.com/a/articles/strategies-to-lengthen-telomeres

https://www.lifeextension.com/
magazine/2016/12/research-update

https://www.frontiersin.org/journals/aging/
articles/10.3389/fragi.2024.1339317/full#B41

References

1. Zhu H, Guo D, Li K, et al. Increased telomerase activity and vitamin D supplementation in overweight African Americans. *Int J Obes (Lond).* 2012;36(6):805-9.

2. Chiappori AA, Kolevska T, Spigel DR, et al. A randomized phase II study of the telomerase inhibitor imetelstat as maintenance therapy for advanced non-small-cell lung cancer. *Ann Oncol.* 2015;26(2):354-62.

3. Cawthon RM, Smith KR, O'Brien E, et al. Association between telomere length in blood and mortality in people aged 60 years or older. *Lancet.* 2003;361(9355):393-5.

4. Xu Q, Parks CG, DeRoo LA, et al. Multivitamin use and telomere length in women. *Am J Clin Nutr.* 2009;89(6):1857-63.

5. Pusceddu I, Herrmann M, Kirsch SH, et al. One-carbon metabolites and telomere length in a prospective and randomized study of B- and/or D-vitamin supplementation. *Eur J Nutr.* 2016.

6. Shin C, Baik I. Leukocyte Telomere Length is Associated With Serum Vitamin B12 and Homocysteine Levels in Older Adults With the Presence of Systemic Inflammation. *Clin Nutr Res.* 2016;5(1):7-14.

7. Min KB, Min JY. Association between leukocyte telomere length and serum carotenoid in US adults. *Eur J Nutr.* 2016.

8. Zhang D, Sun X, Liu J, et al. Homocysteine accelerates senescence of endothelial cells via DNA hypomethylation of human telomerase reverse transcriptase. *Arterioscler Thromb Vasc Biol.* 2015;35(1):71-8.

9. Xiong S, Patrushev N, Forouzandeh F, et al. PGC-1alpha Modulates Telomere Function and DNA Damage in Protecting against Aging-Related Chronic Diseases. *Cell Rep.* 2015;12(9):1391-9.

10. Pusceddu I, Farrell CJ, Di Pierro AM, et al. The role of telomeres and vitamin D in cellular aging and age-related diseases. *Clin Chem Lab Med.* 2015;53(11):1661-78.

11. Harley CB, Liu W, Flom PL, et al. A natural product telomerase activator as part of a health maintenance program: metabolic and cardiovascular response. *Rejuvenation Res.* 2013;16(5):386-95.

12. Borras M, Panizo S, Sarro F, et al. Assessment of the potential role of active vitamin D treatment in telomere length: a case-control study in hemodialysis patients. *Clin Ther.* 2012;34(4):849-56.

13. Makpol S, Zainuddin A, Rahim NA, et al. Alpha-tocopherol modulates hydrogen peroxide-induced DNA damage and telomere shortening of human skin fibroblasts derived from differently aged individuals. *Planta Med.* 2010;76(9):869-75.

14. Tanaka Y, Moritoh Y, Miwa N. Age-dependent telomere-shortening is repressed by phosphorylated alpha-tocopherol together with cellular longevity and intracellular oxidative-stress

reduction in human brain microvascular endotheliocytes. *J Cell Biochem.* 2007;102(3):689-703.

15. Pusceddu I, Herrmann M, Kirsch SH, et al. Prospective study of telomere length and LINE-1 methylation in peripheral blood cells: the role of B vitamins supplementation. *Eur J Nutr.* 2016;55(5):1863-73.

16. Rane G, Koh WP, Kanchi MM, et al. Association Between Leukocyte Telomere Length and Plasma Homocysteine in a Singapore Chinese Population. *Rejuvenation Res.* 2015;18(3):203-10.

17. Bikle DD. Vitamin D: an ancient hormone. *Exp Dermatol.* 2011;20(1):7-13.

18. Gonzalez-Suarez I, Redwood AB, Grotsky DA, et al. A new pathway that regulates 53BP1 stability implicates cathepsin L and vitamin D in DNA repair. *Embo j.* 2011;30(16):3383-96.

19. Liu JJ, Prescott J, Giovannucci E, et al. Plasma vitamin D biomarkers and leukocyte telomere length. *Am J Epidemiol.* 2013;177(12):1411-7.

20. Furumoto K, Inoue E, Nagao N, et al. Age-dependent telomere shortening is slowed down by enrichment of intracellular vitamin C via suppression of oxidative stress. *Life Sci.* 1998;63(11):935-48.

21. Kim YY, Ku SY, Huh Y, et al. Anti-aging effects of vitamin C on human pluripotent stem cell-derived cardiomyocytes. *Age (Dordr).* 2013;35(5):1545-57.

22. Li Y, Zhang W, Chang L, et al. Vitamin C alleviates aging defects in a stem cell model for Werner syndrome. *Protein Cell.* 2016;7(7):478-88.

23. Makpol S, Abidin AZ, Sairin K, et al. gamma-Tocotrienol prevents oxidative stress-induced telomere shortening in human fibroblasts derived from different aged individuals. *Oxid Med Cell Longev.* 2010;3(1):35-43.

24. Makpol S, Durani LW, Chua KH, et al. Tocotrienol-rich fraction prevents cell cycle arrest and elongates telomere length in senescent human diploid fibroblasts. *J Biomed Biotechnol.* 2011;2011:506171.

25. Sen A, Marsche G, Freudenberger P, et al. Association between higher plasma lutein, zeaxanthin, and vitamin C concentrations and longer telomere length: results of the Austrian Stroke Prevention Study. *J Am Geriatr Soc.* 2014;62(2):222-9.

26. Yabuta S, Masaki M, Shidoji Y. Associations of Buccal Cell Telomere Length with Daily Intake of beta-Carotene or alpha-Tocopherol Are Dependent on Carotenoid Metabolism-related Gene Polymorphisms in Healthy Japanese Adults. *J Nutr Health Aging.* 2016;20(3):267-74.

27. Kiecolt-Glaser JK, Epel ES, Belury MA, et al. Omega-3 fatty acids, oxidative stress, and leukocyte telomere length: A randomized controlled trial. *Brain Behav Immun.* 2013;28:16-24.

Chapter – 5

Insulin Resistance and Mitochondrial Dysfunction

The Pancreas, a vital organ, produces the hormone insulin. Insulin is key to maintaining normal blood sugar levels and plays a crucial role in allowing glucose from the bloodstream to enter the body's cells, where it is used for energy. Understanding this process is essential for managing and preventing health issues related to insulin resistance.

However, when the body cannot take the glucose into its cells, extra sugar remains in the bloodstream, and the blood sugar levels are not controlled properly, resulting in type 2 diabetes. This is when the Pancreas becomes exhausted and unable to control the blood sugar levels, or the body becomes immune to insulin, and this is called **Insulin resistance**. Insulin resistance, a condition that demands immediate attention, if left unmanaged, can lead to serious health issues such as type 2 diabetes and heart disease.

Having metabolic syndrome of Insulin Resistance can increase the risk of developing Type 2 diabetes, Heart and blood vessel disease, Hypertension, PCOS, Fatty Liver, and Cancer.

The digestive system breaks down the foods into sugar. Insulin is a hormone made by your Pancreas that lets sugar enter your cells to be used as fuel. In insulin resistance condition, cells don't respond typically to insulin, and glucose can't enter the cells so easily. As a result, your blood sugar levels rise and the Pancreas releases more insulin to try to lower your blood sugar. Insulin resistance is the inability of cells to respond to insulin, which is the main feature of metabolic syndrome that leads to high blood sugar and type 2 diabetes. In simpler terms, it's like the cells in your body are not listening to the insulin's instructions to let glucose in, which can lead to high blood sugar levels.

Insulin resistance is a condition in which the body is not able to use Insulin effectively. As a result, cells become less responsive to Insulin, which leads to cardiovascular disease (CVD), Non-Alcoholic fatty liver, PCOS, and chronic kidney disease (CKD).

Insulin resistance is a condition where the cells in your liver and muscles don't respond effectively to insulin, and it can lead to serious implications for your health. Insulin, a hormone your pancreas produces, is essential for managing your blood glucose. However, when you have insulin resistance, this process is disrupted. Your cells can't effectively absorb glucose, leading to a buildup of blood glucose in your bloodstream.

Despite the cells becoming insulin-resistant, your pancreas continues to produce insulin. This persistence of insulin production is a key factor in the ongoing rise of blood glucose levels.

Insulin, a hormone that regulates the level of blood sugar (Glucose) level in the blood and aids in glucose transportation into the cells, is a key player in our body's metabolic processes.

However, when the body's response to insulin is impaired or the cells cannot effectively utilize insulin, a condition known as Insulin Resistance sets in. This triggers a compensatory response from the pancreas, leading to an overproduction of insulin. This imbalance can have significant health implications, making it crucial to understand the concept of **Insulin Resistance.**

Glucose and lipid metabolism mainly depend on mitochondria to generate energy in cells. Mitochondria are unique as they have their own DNA called mitochondrial DNA (mtDNA). Mutations in this mtDNA or mutations in nuclear DNA (DNA found in the nucleus of a cell) can cause mitochondrial disorder.

Metabolic regulation largely depends on mitochondria, which play an essential role in energy homeostasis by metabolizing nutrients and producing ATP and heat.

An imbalance between energy intake and expenditure leads to mitochondrial dysfunction, characterized by a reduced ratio of energy production (ATP production) to respiration. Genetic and environmental factors, including exercise, diet, aging, and stress, affect both mitochondrial function and insulin sensitivity.

Importantly, it has been shown that mitochondrial dysfunction is associated with insulin resistance in skeletal muscle and other tissues, including the liver, fat, heart, vessels, and pancreas. Thus, insulin resistance caused in part by mitochondrial dysfunction may contribute to a common pathophysiologic etiology for many chronic diseases

Ref from –

https://injection.com/injectable-drugs/26920-insulin-injectable

https://www.ncbi.nlm.nih.gov/pmc/articles/PMC2963150/

https://www.chop.edu/conditions-diseases/mitochondrial-disease#:~:text=Mitochondria%20are%20unique%20in%20[th]at,can%20also%20trigger%20mitochondrial%20disease.

Chapter – 6

CVD and Mitochondrial Dysfunction

Mitochondrial disease, or mitochondrial disorder, refers to a group of disorders that affect the mitochondria, which are tiny compartments in almost every body cell. The mitochondria's primary function is to produce energy. More mitochondria are needed to make more energy, particularly in high-energy-demand organs such as the heart, muscles, and brain. When the number or function of mitochondria in the cell is disrupted, less energy is produced, and organ dysfunction results.

Different symptoms may occur depending on which cells within the body have disrupted mitochondria. Mitochondrial disease can cause a vast array of health concerns, including fatigue, weakness, metabolic strokes, seizures, cardiomyopathy, arrhythmias, developmental or cognitive disabilities, diabetes mellitus, impairment of hearing, vision, growth, liver, gastrointestinal, or kidney function, and more.

These symptoms can present at any age, from infancy until late adulthood. Mitochondrial dysfunction, a feature of heart failure, leads to a progressive decline in bioenergetic reserve capacity. This decline results from a shift of energy production from mitochondrial fatty acid oxidation to glycolytic pathways.

Mitochondrial dysfunction is accountable for the development of cardiovascular diseases (CVDs). Mitochondria produce adenosine triphosphate through oxidative phosphorylation and inevitably generate reactive oxygen species (ROS). Excessive ROS causes mitochondrial dysfunction and cell death.

Mitochondria are semiautonomous organelles located in the cytoplasm. Unlike other organelles, mitochondria have their own genomes (mtDNAs) and double-stranded circular DNAs coated by a double-layered membrane (Davis & Williams, 2012; van der Bliek et al., 2017).

Mitochondria are responsible for producing adenosine triphosphate (ATP) and regulating nutritional metabolism, calcium homeostasis, and cellular viability in several organs (Kim et al., 2008).

Due to the massive demand for ATP in the heart, mitochondria are highly abundant in cardiac cells, especially in cardiomyocytes (CMs), occupying nearly 20–40% of the volume in adult CMs (Schaper et al., 1985). Daily, 6 kg of ATP is produced by cardiac mitochondria (Siasos et al., 2018). Thus, mitochondria play essential roles in the cardiovascular system.

The primary source of energy obtained by oxidative phosphorylation (OXPHOS) comes from the oxidation of fatty acids in the adult heart (Stanley, 2005). ATP is synthesized in adult hearts following the steps below: fatty acyl-coenzyme A (CoAs) are synthesized with the help of acyl CoA syntheses. To enter the cardiac mitochondria, fatty acyl CoAs are converted to acylcarnitines by carnitine palmitoyl transferase 1 (CPT1) and are transferred to the inner membrane of mitochondria (IMM), where they are liberated to fatty acyl CoA and initiate β-oxidation. Acetyl CoA, a product of β-oxidation, enters the Kerbs cycle and generates reduced nicotinamide adenine dinucleotide (NADH) and reduced flavin adenine dinucleotide (FADH2). NADH and FADH2 generated from the Krebs cycle and β-oxidation transfer electrons through the electron transport chain (ETC), which comprises four complexes on the IMM. The efflux of protons accompanied by the electronic flow on ETC activates ATP-synthase and produces ATP (Chen and Butow, 2005).

Ref From –

https://www.chop.edu/conditions-diseases/mitochondrial-disease#:~:text=Mitochondria%20are%20unique%20in%20[th]at,can%20also%20trigger%20mitochondrial%20disease.

https://www.mdpi.com/1422-0067/25/5/2667#:~:text=Mitochondrial%20dysfunction%2C%20a%20feature%20of,acid%20oxidation%20to%20glycolytic%20pathways.

https://www.pamkingsams.com/post/children-s-hospital-of-philadelphia-releases-new-information-on-mitochondrial-disease

https://www.chop.edu/conditions-diseases/mitochondrial-disease

Yang, J., Guo, Q., Feng, X., Liu, Y., & Zhou, Y. (2022). Mitochondrial Dysfunction in Cardiovascular Diseases: Potential Targets for Treatment. Frontiers in Cell and Developmental

Biology. https://doi.org/10.3389/
fcell.2022.841523

https://www.frontiersin.org/journals/cell-
and-developmental-biology/articles/10.3389/
fcell.2022.841523/full

Chapter – 7

Autophagy and Clearance of Pathogens

Autophagy allows your body to break down and reuse old cell parts so your cells can operate more efficiently. It's a natural cleaning-out process that begins when your cells are stressed or deprived of nutrients. Researchers are studying autophagy's role in potentially preventing and fighting disease.

Autophagy is your body's process of reusing old and damaged cell parts. Cells are the basic building blocks of every tissue and organ in your body. Each cell contains multiple parts that keep it functioning. Over time, these parts can become defective or stop working. They become litter, or junk, inside an otherwise healthy cell.

Autophagy is your body's cellular recycling system. Autophagy is also quality control for your cells. Too many junk components in a cell take up space and can slow or prevent a cell from functioning correctly.

Why is autophagy essential?

Autophagy is essential for a cell to survive and do the following functions:

- Recycles damaged cell parts into fully functioning cell parts.

- It removes nonfunctional cell parts that take up space and slow performance.

- Destroys pathogens in a cell that can damage it, like viruses and bacteria.

Autophagy plays an important role when it comes to aging and longevity, too. As a person ages, autophagy decreases, which can lead to a build-up of cellular junk parts and, in turn, cells that aren't functioning at their best.

What happens during autophagy?

Autophagy-related proteins (ATGs) make autophagy possible. ATGs cause structures called autophagosomes to form. Autophagosomes carry the junk cell pieces to a part of the cell called a lysosome. A lysosome's job is to digest or break down other cell parts.

Imagine lysosomes — part of a cell — eating other parts of the cell. The word "autophagy" is a combination of two Greek words translated to mean "self-devouring":

- **"Autos"** means self.

- **"Phagomai"** means to eat.

Lysosomes eat the junk cell parts and release the reusable bits and pieces. The cells use these raw materials to make new parts.

What causes autophagy?

Autophagy occurs when your body's cells are deprived of nutrients or oxygen or if they're damaged in some way.

Think of it this way: Autophagy is a recycling process that makes the most of a cell's already-existing energy resources. The process ramps up when your body has to make the most of these resources because your cells aren't getting them from an outside source.

With autophagy, a cell essentially eats itself to survive. The bonus is that this survival process can lead to cells that work more efficiently.

Can you induce autophagy?

You can induce autophagy by stressing your cells to send them into survival mode. You can induce autophagy through:

- Fasting: You stop eating for a certain amount of time. Fasting deprives your body of nutrients, forcing it to repurpose cell components to function.

- Calorie restriction: Restricting your calories means decreasing the number of energy units, or calories, your body consumes. Instead of completely depriving your body of calories (as with fasting), you limit them. This forces your cells into autophagy to compensate for the lost nutrients.

- Switching to a high-fat, low-carb diet: This type of diet, commonly referred to as a keto diet, changes the way your body burns energy. Instead of burning carbs or sugar for energy, it burns fat. This switch can trigger autophagy.

- Exercise: Exercise stimulates processes that increase the activity of ATGs, such as stressing your skeletal muscles. Exercise can induce autophagy, depending on the type of exercise you're doing and its intensity

Still, being able to induce autophagy doesn't mean you should. For instance, fasting, calorie restriction, or switching to a keto diet may not be safe if you're pregnant, breastfeeding, or if you have a condition like diabetes. Similarly, you shouldn't begin a vigorous exercise routine without consulting a healthcare provider.

Autophagy, an essential biological process that affects immunity, is a powerful tool that host cells can use to defend against infections caused by pathogenic microorganisms.

However, hyperactivated or inhibited autophagy leads to mitochondrial dysfunction, which is harmful to the host and is involved in many types of diseases. Mitochondria perform the functions of biological oxidation and energy exchange. In addition, mitochondrial functions are closely related to cell death, oxygen radical formation, and disease. The accumulation of mitochondrial metabolites affects the survival of intracellular pathogens.

Autophagy is a cellular survival pathway that recycles intracellular components to compensate for nutrient depletion and ensures the appropriate degradation of organelles. Mitochondrial number and health are regulated by mitophagy, a process by which excessive or damaged mitochondria are subjected to autophagic degradation. Autophagy is thus a key determinant for mitochondrial health and proper cell function. Mitophagy malfunction has been recently proposed to contribute to progressive neuronal loss in Parkinson's disease.

In addition to autophagy's significance in mitochondrial integrity, several lines of evidence suggest that mitochondria can also substantially influence the autophagic process.

The mitochondria's ability to influence and be influenced by autophagy places both elements (mitochondria and autophagy) in a unique position where defects in one or the other system could increase the risk of various metabolic and autophagic-related diseases.

Mitochondria, the energy-generating organelles in eukaryotic cells, play essential roles in fundamental cellular processes, such as the supply of metabolic intermediates supporting the cell's biosynthetic and bioenergetic needs, apoptosis, calcium (Ca2 +) homeostasis, and cell signaling.

Mitochondria are the sites of adenosine triphosphate (ATP) production via oxidative phosphorylation (OXPHOS), which takes place in the electron transport chain (ETC). ETC consists of multi-subunit protein complexes embedded in the inner membrane. Electrons are transferred through complexes I–IV to the final electron acceptor oxygen. In the final step, the terminal enzyme of the respiratory chain, cytochrome c oxidase (complex IV) catalyzes the complete reduction of molecular O2 to water.

Mitochondrial Biogenesis

Although it is well established that physical activity increases mitochondrial content in muscle, the molecular mechanisms underlying this process have only recently been elucidated. Mitochondrial dysfunction is an essential component of different diseases associated with aging, such as Type 2 diabetes and Alzheimer's disease. PGC-1α (peroxisome-proliferator-activated receptor γ co-activator-1α) is a co-transcriptional regulation factor that induces mitochondrial biogenesis by activating different transcription factors, including nuclear respiratory factor 1 and nuclear respiratory factor 2, which activate mitochondrial transcription factor A. The latter drives the transcription and replication of mitochondrial DNA. PGC-1α itself is regulated by several different key factors involved in mitochondrial biogenesis.

AMPK acts as the cell's energy sensor and is a key regulator of mitochondrial biogenesis. AMPK activity has been shown to decrease with age, which may contribute to decreased mitochondrial biogenesis and function with aging.

Mitochondrial biogenesis is the process by which a cell synthesizes new mitochondria. It is accomplished by translating nuclear and mitochondrial transcripts and replicating mtDNA. Mitochondrial biogenesis is a mechanism for compensating for mitochondrial damage or responding to metabolic or oxidative stress, and it is driven by the signaling molecule PGC1-α.

Mitochondrial biogenesis is an integral aspect of maintaining mitochondrial quality. In this process, dysfunctional mitochondria are selectively eliminated and replaced by an increase in the number of new mitochondria. Peroxisome proliferator-activated receptor γ coactivator-1α (PGC-1α) is a transcriptional coactivator and a master regulator of mitochondrial biogenesis.

Mitochondrial biogenesis is a major adaptation of skeletal muscle to exercise training. It is induced by a complex interplay between numerous signaling pathways that respond to metabolic, mechanical, and hypoxic stresses generated within the myocyte during contraction. Adipocytes are highly plastic, and mitochondrial biogenesis can be induced by pharmacological interventions in both isolated cells and free-living animals.

Mitochondrial biogenesis is the process through which pre-existing mitochondria grow and divide. It maintains cellular metabolic homeostasis by providing a pool of healthy mitochondria and eliminating damaged mitochondria. Mitochondrial biogenesis requires proteins encoded by both mitochondrial and nuclear genomes.

Autophagy and Mitophagy

Mitochondrial autophagy is a specific autophagy mechanism that removes mitochondria, mainly through the recognition of mitochondrial damage and selective removal of damaged mitochondria. It plays an essential role in maintaining mitochondrial quality control and homeostasis.

Autophagy is an essential lysosome-mediated degradation pathway that maintains cellular homeostasis and viability in response to various intra- and extracellular stresses. Mitophagy is a type of autophagy involved in the intricate removal of dysfunctional mitochondria during conditions of metabolic stress.

Mitochondrial biogenesis and mitophagy are two pathways that regulate mitochondrial content and metabolism, preserving homeostasis. The tight regulation of these opposing processes is essential for cellular adaptation in response to cellular metabolic state, stress, and other intracellular or environmental signals. Interestingly, an imbalance between mitochondrial proliferation and degradation results in the progressive development of numerous pathologic conditions.

Autophagy is the catabolic process through which cells degrade proteins and organelles with a double membrane structure called autophagosomes. It is termed mitophagy when autophagosomes selectively sequester and degrade mitochondria. Together with mitochondrial biogenesis, mitophagy establishes the balance to preserve mitochondrial homeostasis. A regulated mechanism that intentionally eliminates dysfunctional mitochondria is essential for cells to maintain an appropriate bioenergetic system.

Mitophagy has been known to protect cells. It can be induced under stress conditions such as hypoxia, excessive ROS production, or nutrient deprivation.

Mitophagy is a conserved cellular process that is important for the autophagic removal of damaged mitochondria to maintain a healthy mitochondrial population. Mitophagy also appears to occur in plants and has roles in development, stress response, senescence, and programmed cell death.

Mitophagy is a specialized form of autophagy that involves the removal of damaged and surplus mitochondria, depending on the cellular requirements. Mitophagy is triggered by hypoxia, nutrient starvation, mtDNA damage, loss of mitochondrial membrane potential (MMP), and ROS accumulation (Redmann et al., 2014; Hamacher-Brady and Brady, 2016).

Mitophagy can alter the cell state from a highly proliferative cell type with hyperactive mitochondria to a quiescent cell with a limited mitochondrial mass.

Ref From –

https://www.ncbi.nlm.nih.gov/pmc/articles/ PMC3272286/

Ref From - https://www.sciencedirect.com/ science/article/pii/S0005272815000687

https://www.frontiersin.org/journals/cell- and-developmental-biology/articles/10.3389/ fcell.2021.738932/full

https://my.clevelandclinic.org/health/ articles/24058-autophagy

https://www.sciencedirect.com/topics/medicine-and-dentistry/mitochondrial-biogenesis#:~:text=Mitochondrial%20biogenesis%20is%20a%20major,within%20the%20myocyte%20during%20contraction.

https://www.ncbi.nlm.nih.gov/pmc/articles/PMC3883043/

https://www.sciencedirect.com/topics/medicine-and-dentistry/mitochondrial-biogenesis#:~:text=Mitochondrial%20autophagy%20is%20a%20specific,control%20and%20homeostasis%20%5B56%5D.

https://www.sciencedirect.com/science/article/abs/pii/S0531556514000333#:~:text=Mitochondrial%20biogenesis%20and%20mitophagy%20are%20two%20pathways%20that%20regulate%20mitochondrial,other%20intracellular%20or%20environmental%20signals.

https://www.ncbi.nlm.nih.gov/pmc/articles/PMC8466139/

https://www.sciencedirect.com/topics/agricultural-and-biological-sciences/mitophagy

Chapter – 8

Epigenetic changes mitochondria

Mitochondria are vital organelles in the cell that carry out many essential functions necessary for cell survival. These functions include adenosine triphosphate (ATP) production, metal homeostasis, regulation of cellular metabolism, and cellular respiration (Kaniak-Golik and Skoneczna 2015; Lee and Han 2017).

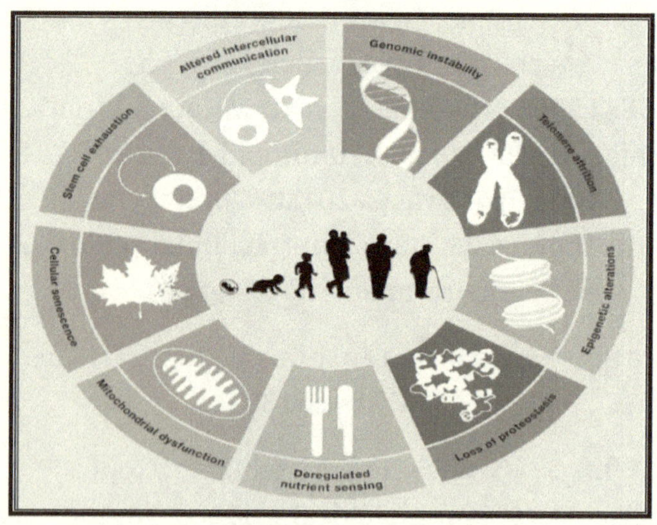

Picture Courtesy From – https://www.ncbi.nlm.nih.gov/pmc/articles/PMC3836174/

https://www.gowinglife.com/what-is-mitochondrial-dysfunction-the-hallmarks-of-ageing-series/

Mitochondria are essential and unique in that they contain their own DNA within the mitochondrial matrix, separate from the genomic nuclear DNA (Yasukawa and Kang 2018).

Mitochondrial dysfunction, resulting from epigenetic changes, in the form of altered gene expression and ATP production, can lead to various conditions, including aging-related neurodegenerative disorders, altered metabolism, changes in circadian rhythm, and cancer.

The expression of genes is known to depend on the genetic sequence and epigenetic regulation of the gene. Epigenetics refers to the study of inheritable changes that would alter gene expression transiently or permanently without permanent modifications to the original DNA sequence.

"Epigenetics" refers to changes in gene expression caused by mechanisms other than changes in the underlying DNA sequences. DNA methylation and histone modification are the two best-studied classic epigenetic regulatory mechanisms that regulate a cell's transcriptional activity in the nucleus.

DNA methylation is a biochemical process where a methyl group is added to the cytosine or adenine, mainly at the C5 position of CpG dinucleotides, by DNA methyltransferase (DNMT).

The mechanisms of epigenetics can be mainly categorized into three classes, namely, DNA methylation, post-translational modifications (PTMs) of histones, and gene expression regulation by non-coding RNAs (ncRNAs). Therefore, Epigenetic mechanisms can be divided into three main categories:

1. Chemical modifications of DNA (i.e., methylation)

2. Post-translational modifications of histone tails

3. Regulation of gene expression by non-coding RNAs

The molecular signatures of epigenetic regulation and chromatin architecture are emerging as pivotal regulators of mitochondrial function. Recent studies unveiled a complex intersection among environmental factors, epigenetic signals, and mitochondrial metabolism, ultimately leading to alterations of vascular phenotype and increased cardiovascular risk.

Changing environmental conditions over the lifetime induce covalent and post-translational chemical modification of the chromatin template, sensitizing the genome to establish new transcriptional programs and, hence, diverse functional states.

On the other hand, metabolic alterations occurring in mitochondria affect the availability of substrates for chromatin-modifying enzymes, thus leading to maladaptive epigenetic signatures altering chromatin accessibility and gene transcription.

Indeed, several components of the epigenetic machinery require intermediates of cellular metabolism (ATP, AcCoA, NADH, α-ketoglutarate) for enzymatic function.

In the present review, we describe the emerging role of epigenetic modifications as fine tuners of gene transcription in mitochondrial dysfunction and vascular disease.

Specifically, the following aspects are described in detail:

1. Mitochondria and vascular function

2. Mitochondrial ROS

3. Epigenetic regulation of mitochondrial function

4. The role of mitochondrial metabolites as key effectors for chromatin-modifying enzymes

5. Epigenetic therapies

Understanding epigenetic routes may lead to new approaches to developing personalized therapies to prevent mitochondrial insufficiency and its complications.

Mitochondria and Vascular Function

Mitochondria, defined as semi-autonomous, membrane-bound organelles localized in the cytoplasm of eukaryotic cells, are emerging as pivotal players in health, disease, and aging. They regulate reactive oxygen species (ROS) production and contribute to retrograde redox signaling from the organelle to the cytosol and nucleus.

Mitochondria are essential in the cellular network formed by metabolic signaling and epigenetic pathways. Indeed, mitochondria drive catabolic and anabolic reactions, supplying energy and metabolites with biosynthetic and signaling roles.

They also maintain a bidirectional signaling crosstalk with the nucleus, generating reciprocal activation-repression (gene expression patterns). Finally, mitochondria can determine apoptotic and necrotic cell death mediated by Ca2+ overload and opening of the permeability transition pore (PTP).

Mitochondria and Type 2 Diabetes Mellitus

Type 2 diabetes mellitus is a major chronic metabolic disorder in public health. Due to mitochondria's indispensable role in the body, its dysfunction has been implicated in the development and progression of multiple diseases, including T2DM. Thus, factors that can regulate mitochondrial function, like mtDNA methylation, are of significant interest in managing T2DM.

Diabetes mellitus (DM) is a chronic metabolic disorder typified by the presence of hyperglycemia. Globally, the prevalence of DM continues to increase by the millions every year.

T2DM is mainly characterized by insulin resistance, with multiple pieces of evidence showing the crucial role of mitochondria involvement. Due to the central role of mitochondria in various cellular responses and signaling pathways, mitochondrial dysfunction will influence the development and progression of T2DM. Thus, information on the association between mitochondria dysfunction and T2DM is of interest to public health. Factors that may regulate mitochondrial dysfunction, such as mitochondrial epigenetics (influenced by external factors like lifestyle intervention), are important to the understanding and treatment of T2DM.

Mitochondrial ROS

Although several cytosolic enzymes (i.e., NADPH, cyclooxygenases, and xanthine oxidase) are implicated in redox balance, ROS generated from mitochondrial oxidative phosphorylation represents the most important source of oxidative stress in vascular cells (i.e., endothelial cells).

Mitochondrial ROS are responsible for the peroxidation of polyunsaturated fatty acids (PUFAs) present in the cellular membrane as well as DNA (causing single and double strand breaks) and protein damage via oxidation of sulfhydryl and aldehyde groups, protein-protein interactions, and fragmentation. In addition, damage to mtDNA may lead to decreased expression of electron transport chain components or defective components that produce more ROS, thus creating a detrimental vicious cycle. mtDNA disruption also correlates with the extent of atherosclerosis in mouse models and human tissues. Despite the highly efficient chemical reduction of O2 through cytochrome *c* oxidase, mitochondria still generate significant levels of ROS. Cellular and mitochondrial physiological levels of ROS are reached when production and scavenging are balanced. Mitochondrial dysfunction is believed to play an essential role in a variety of diseases, including diabetes, obesity, dyslipidaemia, hypertension, arrhythmias, and sudden cardiac death.

Mitochondrial Epigenetics Regulating Inflammation in Cancer and Aging

Inflammation is a defining factor in disease progression; epigenetic modifications of this first line of defense pathway can affect many physiological and pathological conditions, like aging and tumorigenesis. Inflammageing, one of the hallmarks of aging, represents a chronic, low-key, but persistent inflammatory state. Oxidative stress, alterations in mitochondrial DNA (mtDNA) copy number, and mis-localized extra-mitochondrial mtDNA are suggested to induce various immune response pathways directly.

This could ultimately perturb cellular homeostasis and lead to pathological consequences. Epigenetic remodeling of mtDNA by DNA methylation, post-translational modifications of mtDNA binding proteins, and regulation of mitochondrial gene expression by nuclear DNA or mtDNA encoded non-coding RNAs are suggested to directly correlate with the onset and progression of various types of cancer. Mitochondria can also regulate the immune response to multiple infections and tissue damage by producing pro- or anti-inflammatory signals.

This occurs by altering the levels of mitochondrial metabolites and reactive oxygen species (ROS) levels. Since mitochondria are known as the guardians of the inflammatory response, mitochondrial epigenetics might play a pivotal role in inflammation.

Inflammation and Aging

Inflammation, one of the first lines of defense, is frequently repurposed from its fundamental role in immune surveillance to a pro-tumorigenic role. Recent studies report that inflammation can aid the proliferation of cancer cells and promote tumor microenvironment by selectively blocking anti-tumor immunity (Greten and

Grivennikov, 2019). Acute inflammation might be initiated due to several factors, like bacterial or viral infection, autoimmune diseases, obesity, tobacco smoking, asbestos exposure, and excessive alcohol consumption.

Apart from cancers, chronic or acute inflammation is also strongly associated with age-related disorders, including atherosclerosis, diabetes, Alzheimer's disease, rheumatoid arthritis, and aging (Rea et al., 2018).

These epigenetic changes include but are not limited to, alteration in the mitochondrial DNA (mtDNA). Covalent modifications, such as methylation and hydroxymethylation, play a crucial role in altered mtDNA replication and transcription.

DNA Methylation

What happens if I don't methylate well?

- DNA/RNA expression is altered, often leading to chronic diseases (including cancer).

- Neurotransmitter imbalances occur, resulting in several psychological conditions as well as neurodevelopmental delays (including autism spectrum disorder).

- The body's natural detoxification processes are impaired, leading to toxic metabolite buildup and environmental toxins (such as heavy metals).

How does methylation regulation occur?

- The presence or absence of the required vitamins, co-factors, and methyl groups are necessary for the forward movement of the methylation cycle.

- The buildup of any of the substrates within the cycle can and does lead to negative feedback. This, in turn, forces the substrates down alternative pathways, often causing the accumulation of harmful end products.

How does the methylation cycle become impaired?

- Genetic mutations known as SNPs (single nucleotide polymorphisms) within the genes responsible for the production of the needed enzymes can decrease their efficiency, causing a slowdown of the entire cycle.

- Similarly, a lack of the vitamins and co-factors needed results in the same type of decreased enzymatic efficiency.

- Medication interactions

- Environmental toxins such as heavy metals

- Insufficient methyl donors (i.e., methyl folate, methyl B12, and many others)

- High levels of any of the substrate molecules within the cycle

CELL REPAIR

Cell repair requires energy and the raw ingredients needed for RNA to provide the cell with instructions. Impaired methylation leads to decreased responsiveness, which is exactly when the cell needs it the most when under stress. The nervous system, in particular, has the highest concentration of RNA within the body. Therefore, it has the **highest methylation** needs.

CELL SYNTHESIS

Just as with cell repair, cell synthesis requires optimal methylation to keep up with the demand for nucleotides (DNA/RNA building blocks). The human body must create millions of new cells per minute to thrive. Due to their higher demands, tissues such as bone marrow, neural tissue, and red/white blood cells are critically affected by **poor methylation.**

DETOXIFICATION

Many environmental toxins are conjugated to a methyl group before excretion. Poor methylation affects this process, but most importantly, it leads to a decrease in the body's glutathione levels. Glutathione is a highly sulfated protein and a major endogenous antioxidant.

It is also involved in reduction reactions (e.g., maintaining the antioxidants vitamins C and E in their active form); inflammatory control through production and control of leukotrienes; regulation of the nitric oxide cycle; detoxification of heavy metals and other foreign molecules; immune processes such as antigen presentation modulation, cytokine production, lymphocyte proliferation, apoptosis regulation, and enhancement of cytotoxic T cells and NK cells; numerous metabolic/biochemical reactions such as DNA synthesis/repair, amino acid transport, enzyme activation, protein synthesis, and prostaglandin synthesis, etc.

HEART DISEASE

Numerous mutations can occur within the cycle, leading to impaired conversion from the molecule homocysteine to methionine. The resultant increased levels of homocysteine have been demonstrated to be a significant risk factor for cardiovascular disease.

Furthermore, breakdowns within the cycle can also lead to decreased levels of CoQ10, a critical component of energy production on the cellular level. These decreased levels can also result in congestive heart failure and cardiovascular disease. Ironically, the statin drugs often prescribed to lower cholesterol levels have the unfortunate side effect of reducing the body's levels of CoQ10.

ENERGY PRODUCTION

As mentioned with heart disease, **methylation is** essential for producing CoQ10 and mitochondrial fatty acid oxidation, both leading to energy production within cells. When impaired, chronic illnesses such as fibromyalgia, chronic fatigue syndrome, and chronic pain syndromes may occur.

Ref From -

https://www.ncbi.nlm.nih.gov/pmc/articles/ PMC6941438/#:~:text=Mitochondrial%20 dysfunction%20in%20the%20form,in%20 circadian%20rhythm%2C%20and%20cancer.

https://www.frontierspartnerships.org/ articles/10.3389/bjbs.2023.10884/full

https://www.frontiersin.org/journals/
cardiovascular-medicine/articles/10.3389/
fcvm.2020.00028/full

https://www.frontiersin.org/journals/cell-
and-developmental-biology/articles/10.3389/
fcell.2022.929708/full

https://www.sciencedirect.com/science/article/
pii/S2352304215000057

https://www.nbwellness.com/library/
methylation-health-disease/#:~:text=The%20
methylation%20cycle%20is%20
key,dopamine%2C%20serotonin%2C%20and%20
melatonin.

Chapter – 9

Bio Markers of ageing for all the organs

Cardiovascular System

We begin with indicators of cardiovascular functioning, as heart disease is the leading cause of death in the older population and one of the most important causes of disability. The two indicators of blood pressure are probably the most commonly measured biomarkers: *Systolic blood pressure* (SBP) is the maximum pressure in an artery at the moment when the heart is beating and pumping blood; *diastolic blood pressure* (DBP) is the lowest pressure in an artery in the moments between beats when the heart is resting.

Many other biomarkers linked to cardiovascular risk are determined in other ways. One of these is *homocysteine*, an amino acid measured from blood plasma. Homocysteine affects atherosclerosis development by damaging the arteries' inner lining and promoting blood clots.

Homocysteine has garnered recent attention because of its importance in predicting many of the major health outcomes expected in aging populations, including cardiovascular disease, peripheral vascular disease, and poorer cognitive function. It is highly related to dietary content including folate and vitamins B12 and B6.

The next set of markers is indicators of metabolic processes, many of which are also related to cardiovascular outcomes. *Cholesterol* has several functions including keeping cell membranes intact and helping the synthesis of steroid hormone and bile acids.

In recent years, components of total cholesterol are generally measured to determine the risk for heart disease: *low-density lipoprotein* (LDL), *high-density lipoprotein* (HDL), and *very low-density lipoprotein* (VLDL). High levels of HDL are protective for heart disease because HDL carries cholesterol away from the arteries and back to the liver, where it is passed from the body. Thus, HDL is called the "good" cholesterol, and low levels are associated with higher risk.

Triglycerides, an indicator of stored fat, are often included among the lipid indicators in an evaluation of coronary risk factors.

Fasting blood glucose level is indicative of diabetes and prediabetes. Higher than average blood glucose contributes to the development of metabolic syndrome.

Markers of Inflammation, Immunity, and Infection

Markers of Inflammation are the following category of markers. Age-related changes in inflammatory markers are complex and include a wide range of potential indicators. Here, we focus on the markers most commonly used in aging research. *C-reactive protein* (CRP) is an acute phase response protein produced in the liver that indicates general systemic levels of inflammation. CRP levels rise as part of the immune response to infection and tissue damage or injury and may be elevated due to the presence of chronic conditions like diabetes, asthma, rheumatoid arthritis, and heart disease.

Markers of Activity in the Hypothalamic Pituitary Axis

Cortisol is a steroid hormone the adrenal cortex produces in response to internal or external stress. Consistently high cortisol reactivity to repeated challenges is an atypical response that may reflect chronic physiological stress and is associated with adverse health outcomes in old age.

Cortisol levels have been shown to be greater among individuals experiencing chronic stress from work or emotional strain. Health consequences of exposure to elevated cortisol include increased cardiovascular risk and poorer cognitive functioning.

Markers of Oxidative Stress and Antioxidants

Oxidative stress and antioxidants are examples of markers that seem theoretically essential determinants of the aging process but are as yet not measured in such a way that they can be collected from large populations. High levels of reactive oxidative species (ROS), enzymes important in cell signaling, have been shown to cause significant damage to cell structures. It has been suggested that ROS plays an important role in the onset of age-associated loss in muscle mass (sarcopenia), changes in the central nervous system, hearing loss, Parkinson's disease.

Genetic Markers

Mutations in mitochondrial DNA (mtDNA) *accumulate with age and are among the genetic factors that may eventually be associated with longevity.*

Telomere length is another genetic indicator currently under investigation as either an indicator of the risk of aging or a biological marker of the aging process per se.

Ref From –

https://www.ncbi.nlm.nih.gov/pmc/articles/
PMC5938178/

Crimmins, E. M., Vasunilashorn, S., Kim, J. K., & Alley, D. E. (2008). Chapter 5 Biomarkers Related To Aging In Human Populations. Advances in Clinical Chemistry. https://doi.org/10.1016/s0065-2423(08)00405-8

Blanchflower, D. G., & Bryson, A. (2024). The adult consequences of being bullied in childhood. https://doi.org/10.1016/j.socscimed.2024.116690

Andrews, V. H., & Borkovec, T. D. (1988). The differential effects of inductions of worry, somatic anxiety, and depression on emotional experience. Journal of Behavior Therapy and Experimental Psychiatry. https://doi.org/10.1016/0005-7916(88)90006-7

Patlevič, P., Vašková, J., Švorc, P., Vaško, L., & Švorc, P. (2016). Reactive oxygen species and antioxidant defense in human gastrointestinal diseases. https://doi.org/10.1016/j.imr.2016.07.004

Chapter – 10

What Is Methylation?

Methylation is a critical biochemical pathway that occurs in our bodies. It occurs billions of times every second within our cells.

Without methylation, we would die. Methylation is crucial in many bodily functions, including detoxification, immune function, energy production, mood balancing, DNA maintenance, and inflammation management.

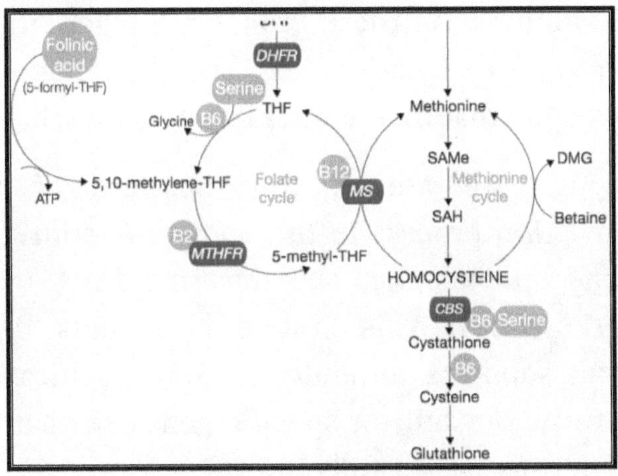

Picture Courtesy From - https://www.sydneynaturalfertility.com/the-mthfr-gene-and-pregnancy-infertility-and-miscarriage/

What Is MTHFR?

MTHFR is a gene that creates an enzyme called **methylenetetrahydrofolate reductase** (MTHFR). This gene/enzyme function creates a vital multi-step chemical breakdown/cycling process called methylation.

This highly intricate process of methylation occurs in every cell in the body and the fluid supplying the brain.

Methylation Process

This methylation process is like a factory line where '**methyl groups**' are created. These methyl groups comprise a carbon with three hydrogen (CH3) atoms. As these groups are made 'in the factory' of methylation, it is passed onto another 'worker' so that that worker can do its work.

Although we are entirely unaware of this methylation process in the body, *it is critical for making, maintaining, and repairing DNA (your genetic code)*. This system also shuts down viruses, supports immunity, assists detoxification, and turns off and on specific genes so that the system runs successfully.

This is known as **gene expression**. Think of it this way- methylation is like a factory line where car tires are being made. The car (without the tires) is waiting on the lot to receive its tires so it can drive off the lot to go do its job. If tires (methyl groups) aren't being made in the factory, the car can't do what it was created to do.

Specific Jobs of Methylation

More specific jobs of methylation include:

- Processing and filtering heavy metals and environmental toxins out of the body

- Homocysteine management to prevent the harmful effects of high homocysteine (inflammation in the blood vessels, brain, and heart tissue, and increased risk of developing Alzheimer's)

- Converting nutrients like folic acid to methyltetrahydrofolate so that it can do innumerable actions throughout the body

- Nervous system function and insulation around nerves

- Creating the building blocks, RNA and DNA

- Efficient removal of fat and cholesterol or appropriately used to create hormones

- Regulate gene expression

- Immune function - removal of free radicals

- Manages the amount of fluid inside and outside of each cell

- Neurotransmitter and chemical messenger communication and processing

- Manages sulfur metabolism, which is crucial for the detoxification process

- The production and repair of proteins in the body

- Histamine regulation and allergic responses

- Waste and toxin cleanup through BH4 production

Biochemistry

Here's a bit of **biochemistry** for your enjoyment.

There are four major molecules involved in the methylation cycle:

- Homocysteine

- Methionine

- S-adenosylmethionine (SAMe)

- S-adenosylhomocysteine.

Each molecule is like a step in the factory line, and you get from one molecule to the next by means of enzymes. Each enzyme has an important co-enzyme or cofactor that it relies on to function properly and continue on to the next step.

These cofactors are critical to enzyme functioning properly, and **they include B vitamins such as folic acid, vitamin B12, and vitamin B.**

A biochemical glitch happens in this cycle. For this 'factory line' of methylation to work, these **B vitamins** must be in active form. **Vitamin B12 needs to be in the form methylcobalamin; folic acid needs to be form folinic acid and pyridoxyl-5-phosphate**

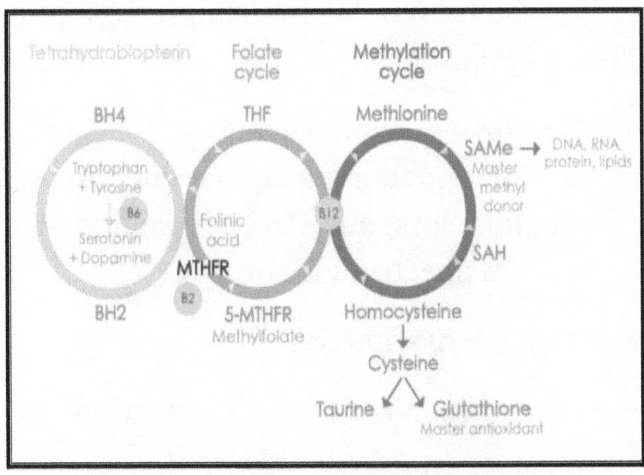

Pic Courtesy From - https://medium.com/@shealarroque/mthfr-genetics-5747d04bb00c

The process of methylation produces the most **active form of folate in your body**, known as **methylfolate**. Methionine is essential and is used by the body to help detoxify against harmful agents such as lead and other heavy metals; it supplies sulfur and other compounds required by the body for normal metabolism and growth, it reduces the levels of histamine, and it has been found to act like an antioxidant to remove free radicals from the body, and it promotes excretion of estrogen.

MTHFR Gene Mutation Explained

What happens when your MTHFR gene is defective/mutated? This is called an SNP ('snip') which stands for **single nucleotide polymorphism**.

- Neurologic abnormalities - such as weakness, numbness, fatigue, seizures, memory loss, and likely increased risk of Autism Spectrum Disorders.

- Immune dysfunction.

- Poor protein repair and production

- Neuropsychiatric imbalances leading to depression, anxiety, panic attacks, irritability, mood swings, OCD, ADD/

ADHD, Parkinson's, schizophrenia, bipolar disorder, and poor concentration.

- Lack of hormone regulation.

- Poor cell turnover, repair, and maintenance.

- Increased risk of cancer through a couple of different mechanisms.

- Inappropriate genetic coding leading to additional genetic mutations.

- Cholesterol accumulation leading to fat deposits around organs and within blood vessels.

- Significantly decreased glutathione production. Glutathione is the body's most potent detoxifier and without it toxins accumulate.

- Increased risk for stroke, heart disease, clotting disorders, infertility, spina bifida, and many chronic diseases.

- Heavy metal toxicity and toxic substance accumulation in the body.

- Electrolyte imbalances and cellular irregularities leading to Alzheimer's,

Autism, Multiple Sclerosis, and Crohn's Disease.

- Abnormal DNA and RNA function.

- Increased risk of autoimmune disease.

- Hormone abnormalities leading to diabetes as well as pituitary, thyroid, and adrenal dysfunction.

- Widespread cellular damage leading to headaches, fatigue, intestinal permeability (leaky gut), and many other inflammatory conditions.

Folic Acid vs. Folate — What's the Difference?

Folate and folic acid are different forms of vitamin B9. While there's a distinct difference between the two, their names are often used interchangeably.

B Complex and Vitamin B9 (Folate)

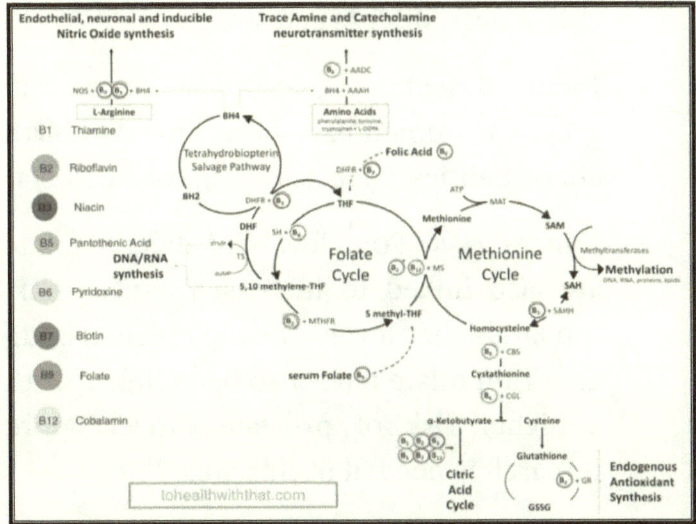

Picture Courtesy From - https://www.tohealthwiththat.com/what-else-to-take-for-mthfr/

Vitamin B9 is an essential nutrient that naturally occurs as folate.

It serves many vital functions in your body. For example, it plays a crucial role in cell growth and DNA formation.

Low levels of vitamin B9 are associated with an increased risk of several health conditions, including:

- **Elevated homocysteine.** High homocysteine levels have been associated with an increased risk of heart disease and stroke.

- **Birth defects.** Low folate levels in pregnant women have been linked to birth abnormalities, such as neural tube defects.

- **Cancer risk.** Poor levels of dietary folate are also linked to increased cancer risk, though higher levels of supplemental and/or serum folate have also been linked with a higher risk of prostate cancer. More research is needed in this area.

What is folate?

Folate is the naturally occurring form of vitamin B9. Its name is derived from the Latin word "folium," which means leaf. Leafy vegetables are among the best dietary sources of folate.

Folate is a generic name for a group of related compounds with similar nutritional properties. The active form of vitamin B9 is a type of folate known as 5-methyltetrahydrofolate (5-MTHF). Before entering your bloodstream, your digestive system converts folate to the biologically active vitamin B9 — 5-MTHF form. Most dietary folate is converted into 5-MTHF in your digestive system before entering your bloodstream.

What is folic acid?

Folic acid is a synthetic form of vitamin B9 known as monopteroylglutamic acid or pteroylmonoglutamic acid.

It's used in supplements and added to processed food products, such as flour and breakfast cereals.

Unlike folate, not all of the folic acid you consume is converted into the active form of vitamin B9 — 5-MTHF — in your digestive system. Instead, some folic acid is converted to 5-MTHF in your liver.

Yet, this process is slow and inefficient for some people. After taking a folic acid supplement, it takes time for your body to convert all of it to 5-MTHF.

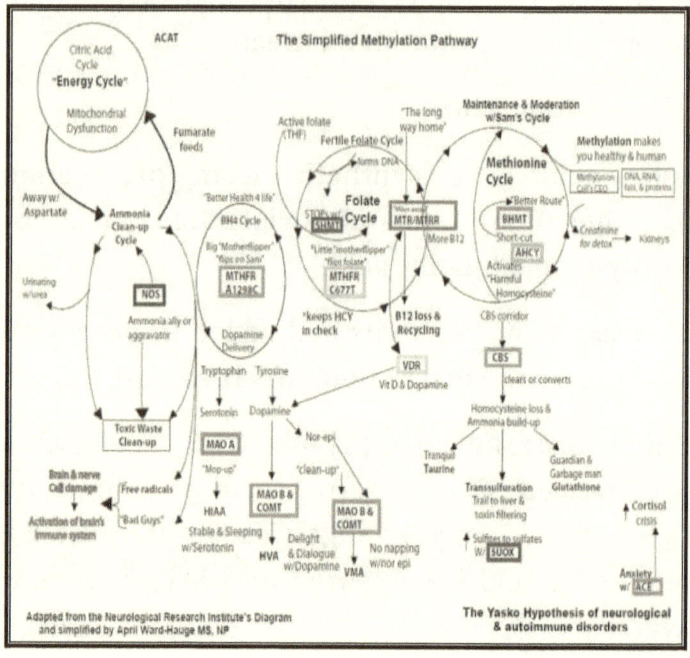

Pic Courtesy From - https://www.inspireyourhealth.com/understanding-mthfr/

Ref From –

Folic Acid vs. Folate – Know Your Iron! – Brides & You. https://bridesandyou.com/2022/12/15/folic-acid-vs-folate-know-your-iron/

Folic Acid vs. Folate — What's the Difference? https://www.healthline.com/nutrition/folic-acid-vs-folate

https://fullyfunctional.com/blog/what-is-mthfr-all-you-need-to-know-about-methylation-and-mthfr/#:~:text=MTHFR%20is%20a%20gene%20that,breakdown%2Fcycling%20process%20called%20methylation.

Chapter – 11

Cell Injury

Apoptosis, Necrosis, Autophagy, and Senescence

Cell damage and Mitochondria

Aging is associated with increased accumulation of cellular damage. Many types of stress can cause cellular damage, including low oxygen levels, DNA alterations, low nutrient levels, and oxidative stress (exposure to increased levels of reactive oxygen species (ROS)).

Damage from these stressors can include DNA mutations, protein unfolding, and oxidation of lipids in membranes, all of which can impair cellular function. Because mitochondria are sites of high ROS production, they are vital organelles in aging.

Mitochondria also are essential organelles in the induction of apoptosis, autophagy, and senescence.

Cell death, like apoptosis and autophagy, are natural processes. Cells die or recycle to make way for newer, more efficient ones. Your body expects these programmed cell deaths to happen.

Necrosis (unexpected cell death from lack of blood flow) causes tissue death. When cell death doesn't happen as it should, cancers and other problems occur.

What is cell death?

Cell death occurs when cells in your body stop working and die. Cells in your body reproduce — a process called cell division or mitosis. Experts believe healthy human cells can replicate or divide up to 60 times before cell death occurs.

Your body is constantly making new cells to replace damaged and dying ones. Natural cell death keeps your body healthy and functioning. Problems arise when cell death doesn't happen as expected, or cells die when they shouldn't.

What are cells?

Cells are the structures that make up all living forms. In fact, your body has more than 30 trillion cells. Cells have three main parts:

1. **Cell membrane: This thin layer surrounds a cell. It has receptors that control what goes in and out of the cell.**

2. **Cytoplasm: This fluid inside each cell makes proteins. It's also where most chemical reactions inside a cell take place.**

3. **Nucleus: This structure inside each cell contains most of your genetic information in the form of deoxyribonucleic acid (DNA). The nucleus also makes ribonucleic acid (RNA). RNA carries DNA, which develops into proteins that cells need to function.**

What causes cell death?

- There are many reasons cells die:

- Some cells die as they develop before they're fully formed.

- Old cells reach the age where they can't divide anymore and die.

- Irreparable damaged cells naturally die off.

- Diseases, injuries, toxins, and specific treatments damage cells, causing cell death.

Who is at risk for cell death?

Cell death affects all of us — and that's usually a good thing. Your skin sheds up to 40,000 dead cells every day. Your skin's outer layer, the epidermis, constantly makes new skin cells to replace them.

In other words, you wouldn't have skin without this cell death.

What are the types of cell death?

There are three main types of cell death:

- Apoptosis.

- Autophagy.

- Necrosis.

What is apoptosis and programmed cell death?

Apoptosis is a form of programmed cell death that happens when cells naturally self-destruct or die. With apoptosis or programmed cell death, cells die when they achieve maximum cell division and can no longer reproduce. This type of programmed cell death is healthy and expected.

An example of apoptosis occurs during pregnancy. Unnecessary cells between developing fetal fingers die. After cell death, the fingers can separate. If apoptosis doesn't happen, a baby may be born with skin webbing between their fingers that fuses them together. This congenital hand difference is called syndactyly.

Apoptosis is an ongoing process. It rids your body of old, damaged cells so that younger, healthier cells can take their place. When something stops this programmed cell death from happening, old and damaged cells can uncontrollably reproduce.

This is how tumors and cancers form. Sometimes, there are problems with cell programming, and cell death occurs when it shouldn't. When you develop Parkinson's disease, Huntington's disease, or Alzheimer's disease, too many nerve cells or neurons in your nervous system die. This premature cell death affects your ability to think or move.

What is autophagy?

Autophagy is another form of programmed cell death. It typically occurs during times of stress or hunger. It's your body's way of breaking down and reusing old cell parts to make more efficient cells. You might think of autophagy as your body's cellular recycling system.

During autophagy (self-devouring), a cell consumes old or damaged proteins and other substances in its cytoplasm. The cell then recycles these broken-down parts to support essential cell functions.

Autophagy can be helpful when the recycling process assists your immune system in destroying infection-causing viruses and bacteria. It can also prevent healthy cells from becoming cancerous and may prevent other problems like heart disease.

On the other hand, autophagy can be impaired if you have cancer. The recycling process provides extra nutrients to cancer cells, fueling their growth. Plus, the recycled cells may prevent cancer treatments from destroying the cancer cells.

Some people severely restrict calories or fast to induce autophagy and stimulate the production of younger, healthier cells. However, there's no evidence that this type of induced autophagy is successful. We do know that strict dietary restrictions can be unhealthy.

What is necrosis?

Necrosis is an accidental or unprogrammed cell death that causes tissue death. Trauma to a cell can cause its contents to leak and damage nearby cells. This can lead to inflammation and additional cell damage, and death. Although your healthcare provider can remove the dead tissue, tissue death is irreversible.

Lack of blood flow and oxygen to certain areas of your body can cause cell death. Necrosis also happens when you die because cells no longer receive blood, oxygen and nutrients.

Other causes of necrosis include:

- Accidents and traumatic injuries.

- Autoimmune diseases.

- Infection-causing bacteria, viruses and fungi.

- Poisons, toxins and illicit drug use.

- Radiation therapy.

Types of necrosis

Types of necrosis vary depending on the cause and affected body area. Common necrosis types include:

- Avascular necrosis or osteonecrosis (bone tissue death).

- Gangrene (skin tissue death).

- Pulp necrosis (tooth death).

References

- Alberts B, Johnson A, Lewis J, et al. Programmed Cell Death (Apoptosis). In: *Molecular Biology of the Cell (https://www.*

ncbi.nlm.nih.gov/books/NBK26873/).   4th edition. New York: Garland Science; 2002. Accessed 9/28/2023.

- Galluzzi L, Vitale I, Aaronson S, et al. Molecular Mechanisms of Cell Death: Recommendations of the Nomenclature Committee on Cell Death 2018 *(https:// www.ncbi.nlm.nih.gov/pmc/articles/ PMC5864239/)*. *Cell Death Differ.* 2018 Mar; 25(3):486-541. Accessed 9/28/2023.

- Khalid N, Azimpouran M. Necrosis *(https://www.ncbi.nlm.nih.gov/books/ NBK557627/)*. 2023 Mar 6. In: StatPearls [Internet]. Treasure Island (FL): StatPearls Publishing. Accessed 9/28/2023.

- Liu X, Yang W, Guan Z, et al. There are only four basic modes of cell death, although there are many ad-hoc variants adapted to different situations *(https://www.ncbi. nlm.nih.gov/pmc/articles/PMC5796572/)*. *Cell Biosci.* 2018 Feb 1;8:6. Accessed 9/28/2023.

- National Cancer Institute (U.S.). Autophagy *(https://www.cancer.gov/publications/ dictionaries/cancer-terms/def/autophagy)*. Accessed 9/28/2023.

- National Cancer Institute (U.S.). Cell (*https://www.cancer.gov/publications/ dictionaries/cancer-terms/def/cell*). Accessed 9/28/2023.

- National Cancer Institute (U.S.). Programmed Cell Death (*https://www. cancer.gov/publications/dictionaries/ cancer-terms/def/programmed-cell-death*). Accessed 9/28/2023.

- National Human Genome Research Institute (U.S.). Apoptosis (*https://www. genome.gov/genetics-glossary/apoptosis*). Accessed 9/28/2023.

- Yan G, Elbadawi M, Efferth T. Multiple cell death modalities and their key features (Review) (*https://pdfs.semanticscholar. org/36f2/73583395 66f8d 684da6a17bfbdf626a373d0.pdf?_ gl=1*7wnfcm*_ga*MTc1MDIwNzY3M S4xNjk2NTM2NzI5*_ga_ H7P4ZT52H5*MTY5NjUzNjcyOC4xLjAu MTY5NjUzNjczMS41Ny4wLjA.*). World Acad Sci J. 2020 Mar;2:39-48. Accessed 9/28/2023.

Ref From –

https://now.aapmr.org/cell-death-apoptosis/

https://my.clevelandclinic.org/health/articles/
cell-death

Chapter – 12

Blood-Brain Barrier

Your blood-brain barrier (BBB) is a tightly locked layer of cells that defend your brain from harmful substances, germs, and other things that could cause damage. It's a vital part of maintaining your brain health. It also holds good things inside your brain, maintaining the organ's delicate chemical balance.

Your blood-brain barrier (BBB) is a protective layer that lines the inner surfaces of the blood vessels inside your brain. It's a vital part of how your brain and nervous system work.

Though the name includes the word "barrier," it's more of a filter and behaves like a gatekeeper to your brain. It's there to keep harmful things out and hold helpful things in. It also controls how various chemical molecules (including compounds your body needs or makes itself) enter and exit your brain.

The blood-brain barrier (BBB) is a selective semi-permeable membrane between the blood and the interstitium of the brain. It allows cerebral blood vessels to regulate molecule and ion movement between the blood and the brain.

The BBB is composed of endothelial cells (ECs), pericytes (PCs), capillary basement membrane, and astrocyte end-feet, all of which aim to shield the brain from toxic substances, filter harmful compounds from the brain to the bloodstream, and supply brain tissue with nutrients.

The BBB has physical (tight junctions) and metabolic (enzyme) barriers to do this. Central nervous system (CNS) structures are unique in structure and function and, therefore, require a stable environment with a composition that differs from that of the peripheral circulation. For this reason, the BBB exists to maintain a homeostatic environment in which CNS structures can function without disruption from other bodily functions.

The BBB is responsible for creating and maintaining homeostasis for neuronal functions, defending the system against toxic insults, regulating the communication between the periphery and the CNS, and providing the brain with nutrients.

This is achieved via four main mechanisms:

- prevention of the paracellular diffusion of hydrophilic compounds

- mediation of the active transport of nutrients to the brain

- activation of efflux transport of hydrophobic molecules and drugs from the brain to the blood

- regulation of the trans endothelial migration of circulating blood cells and pathogens

What does the blood-brain barrier do?

Inside your blood vessels is a layer of specialized cells called the endothelium. But inside your brain, the endothelium is different.

Endothelial cells lining the inside of your brain's blood vessels are tightly packed together, forming your blood-brain barrier. They're so tightly packed that there's almost no space for anything to slip through without help. These cells have a lipid-based outer membrane.

What can get through the BBB?

Some things can get through your BBB if they're small enough. Others can get through because they're lipid-soluble. That means they can pass through your blood-brain barrier without it repelling them. Larger or water-soluble molecules can't get through the BBB on their own.

Large molecules can't slip between the interlocking endothelial cells because of their size. Water-soluble molecules can't easily pass through your BBB because its cell membranes are lipid-based, which repels water-soluble molecules.

If large or water-soluble molecules — including nutrients — need to get through, they need transportation to help them across.

Some examples of drugs and substances that can get through the BBB (either on their own or with transport help) include:

Alcohol.

Anesthetics.

Antidepressant medications.

Anxiolytics (antianxiety medications).

Antipsychotic medications.

Medications that treat seizures or epilepsy.

Caffeine.

Acetaminophen and most nonsteroidal anti-inflammatory drugs (NSAIDs).

Sedative hypnotics (such as barbiturates, benzodiazepines and similar drugs).

The above list mainly refers broadly to medication classes that can pass through your BBB. However, medications from many more classes can also make it through.

The list of what can make it through your BBB is exceptionally lengthy, so medications that fall into the above classes make up only a fraction of the medications that can cross your BBB.

Currently, the best methods to find or predict which compounds can pass through your BBB involve highly complex algorithms and computer programming. So far, these methods list nearly 5,000 chemical molecules (including medications) that can (or should be able to) pass through your BBB.

What are the common conditions and disorders that affect the blood-brain barrier?

Your blood-brain barrier is highly secure, but it isn't perfect. Inflammation can weaken it. Other conditions can also make it less effective. That can let harmful substances or pathogens into your brain.

What can I do to take care of the health of my blood-brain barrier?

There's no direct way to prevent disruptions that affect your BBB. Instead, the best approach is to avoid conditions or circumstances that could lead to its disruption. Some things you can do include:

- **Eat a balanced diet and maintain a healthy weight.** Conditions affecting circulatory health, especially stroke, can disrupt your blood-brain barrier. Reaching and maintaining a weight that's healthy for you and following dietary guidance from your primary care provider are very important. They can help you prevent — or at least delay or reduce the severity of — a stroke. That can help protect your BBB.

- **Don't ignore infections.** Eye and ear infections can spread and cause inflammation that disrupts your BBB. That can let them access your brain, which can be serious or deadly.

- **Manage your health conditions**. Chronic conditions like Type 2 diabetes, high blood pressure, high cholesterol and epilepsy can disrupt your BBB and the blood vessels it surrounds. Managing these conditions, with either medication or other treatments, is an important step in protecting your blood-brain barrier.

Are there any common treatments that target the blood-brain barrier?

For now, there aren't any treatments that target your blood-brain barrier directly. Instead, treatments target conditions that could disrupt your BBB. Some examples include:

- Medications for high blood pressure, high cholesterol or high blood sugar.

- Treatments to restore blood circulation in your brain after an ischemic stroke (this is a stroke that involves a blockage that cuts off blood flow to part of your brain).

- Medications that prevent seizures.

- Treatments for brain cancer.

Conditions and Disorders

What are the common conditions and disorders that affect the blood-brain barrier?

Your blood-brain barrier is highly secure, but it isn't perfect. Inflammation can weaken it. Other conditions can also make it less effective. That can let harmful substances or pathogens into your brain (it can also make some medications, such as penicillin-type antibiotics, more effective at treating infections in your brain).

Conditions that involve any weakening in your BBB's integrity include both acute and chronic illnesses.

Acute conditions involving weakened BBB integrity

Acute conditions are issues that are happening right now or very recently. Examples of acute conditions that can affect your BBB include (but aren't limited to):

Brain cancer.

Brain infections (encephalitis and meningitis).

Concussion and traumatic brain injury (TBI).

High blood carbon dioxide levels (hypercapnia).

Low blood oxygen levels (cerebral hypoxia).

Stroke.

Seizures (especially ones that turn into status epilepticus).

Chronic conditions involving weakened BBB integrity

Chronic conditions are long-term concerns. They can last months (at minimum), but many — if not most — are permanent and can last for years. These include (but aren't limited to):

Alzheimer's disease.

Amyotrophic lateral sclerosis (ALS).

Chronic hypercapnia from conditions like chronic obstructive pulmonary disease (COPD).

Epilepsy.

Frontotemporal dementia.

High blood pressure (hypertension).

High blood sugar (hyperglycaemia) and related conditions like Type 2 diabetes.

High cholesterol (hyperlipidaemia).

Multiple sclerosis.

Neuromyelitis optica (NMO).

Parkinson's disease.

Experts suspect many other chronic conditions can affect your BBB, but more research is necessary to confirm this.

Ketone Bodies Reaching the Brain

Under normal physiological conditions, the brain primarily utilizes glucose for ATP generation. However, in situations where glucose is sparse, e.g., during prolonged fasting, ketone bodies become an essential energy source for the brain. The brain's utilization of ketones seems to depend mainly on the concentration in the blood. Thus, many dietary approaches, such as ketogenic diets, ingesting ketogenic medium-chain fatty acids, or exogenous ketones, facilitate significant brain metabolism changes.

While the brain primarily relies on glucose as the primary fuel, other substrates may contribute to metabolism, mainly when glucose supply is restricted or inadequate, e.g., during fasting and low carbohydrate diets. Ketone bodies, together with lactate, are the primary alternative fuels for the brain, and both are able to cross the blood-brain barrier through monocarboxylate transporters (MCTs) in endothelial cells and astroglia.

Plasma ketone levels are usually low after an overnight fast (<0.5 mM) and contribute to less than 5% of the brain's metabolism. However, during prolonged fasting (5–6 weeks), ketone body levels rise significantly and can contribute almost 60% of the brain's energy requirement, thereby replacing glucose as the primary fuel. Ketonemia can be achieved in non-fasting states by ketogenic diets or by the ingestion of supplements in the form of ketogenic medium-chain fatty acids (MCFA) or exogenous ketone esters or salts.

What is ketosis?

Ketosis occurs when the body uses fat as its main fuel source. Normally, the body uses blood sugar (glucose) as its key energy source.

You typically get glucose in your diet by eating carbohydrates (carbs) such as starches and sugars. Your body breaks the carbohydrates into glucose and then uses the glucose as fuel. Your liver stores the rest and releases it as needed.

When your carb intake is very low, these glucose stores drain down. Since your body doesn't have enough carbs to burn for energy, it burns fat instead. As your body breaks down fat, it produces a compound called ketones. The ketones, or ketone bodies, become your body and brain's main energy source.

The fat your body uses to create ketones may come from your diet (nutritional ketosis), or it may come from your body's fat stores. Your liver produces a small amount of ketones on its own. But when your glucose level decreases, your insulin level decreases. This causes your liver to ramp up the production of ketones to ensure it can provide enough energy for your brain. Therefore, your blood has high levels of ketones during ketosis.

What is the ketosis diet?

The ketogenic (keto) diet changes the way your body uses food. Typically, carbohydrates in your diet provide most of the fuel your body needs. The keto diet reduces the number of carbs you eat and teaches your body to burn fat for fuel instead.

The keto diet is high in fat, moderate in protein, and low in carbohydrates. The standard keto diet consists of 70% to 80% fat, 10% to 20% proteins, and 5% to 10% carbohydrates.

Many nutrient-rich foods contain high amounts of carbohydrates. This includes whole grains, fruits and vegetables. Carbs from all sources are restricted on the keto diet. So, you'll have to cut out all bread, cereal, and other grains and make serious cuts to your fruit and vegetable intake.

The types of foods that provide fat for the keto diet include:

Meats and fish.

Eggs.

Nuts and seeds.

Butter and cream.

Cheese.

Oils such as olive oil and canola oil.

If you eat a high-carb diet before starting a keto diet, it may take you longer to reach ketosis than someone who consumes a low-carb diet. That's because your body needs to exhaust its glucose stores first.

You may be able to get into ketosis faster with intermittent fasting. The most common method of intermittent fasting involves eating all of your food within eight hours. Then, you

fast for the remaining 16 hours of a 24-hour period.

What are the benefits of ketosis?

Research has shown that ketosis may have several health benefits. One of the biggest benefits of ketosis may be weight loss. The process can help you feel less hungry, which may lead to eating less food. It can help you lose belly fat (visceral fat) while maintaining a lean mass. Other possible benefits of ketosis include treating and managing diseases such as:

Epilepsy: Healthcare providers often put children with epilepsy on the keto diet to reduce or even prevent seizures by altering the "excitability" part of their brain.

Other neurologic conditions: Research has shown the keto diet may help improve neurological conditions such as Alzheimer's disease, autism and brain cancers such as glioblastoma.

Type 2 diabetes: The keto diet can help people with Type 2 diabetes lose weight and manage their blood sugar levels.

Heart disease: The keto diet may lower your risk of developing cardiovascular disease by lowering your blood pressure, improving your HDL ("good") cholesterol levels and lowering your triglycerides.

Metabolic syndrome: The keto diet may reduce your risk of developing metabolic syndrome, which is associated with your risk of heart disease.

Ketosis has also been shown to increase your focus and energy. The keto diet delivers your body's energy needs in a way that reduces inflammation. Research suggests your brain works more efficiently on ketones than on glucose.

What are the side effects of ketosis?

The keto diet has many benefits, but it may come with some side effects. One of the signs of ketosis may include "keto flu," which includes symptoms such as upset stomach, headache and fatigue.

Other symptoms of ketosis may include:

Bad breath ("keto" breath).

Constipation.

Insomnia.

Dehydration.

Low bone density (osteopenia) and bone fractures.

High cholesterol (hyperlipidemia).

Kidney stones.

What's the difference between ketosis and diabetes-related ketoacidosis (DKA)?

Ketosis and diabetes-related ketoacidosis (DKA) are two very different things. During ketosis, you have ketones in your blood but not enough to turn your blood acidic. It usually happens if you're fasting or following a low-carbohydrate diet. Ketosis isn't harmful.

DKA is a condition that affects people with diabetes and people with undiagnosed diabetes. It happens when your blood turns acidic because it has too many ketones due to a lack of insulin. Diabetes-related ketoacidosis is life-threatening and requires immediate medical attention.

A note from Cleveland Clinic

Ketosis is a metabolic state that occurs when your body burns fat for energy instead of glucose. The keto diet has many possible benefits. These include potential weight loss, increased energy and treating chronic illness. However, the diet can be difficult to follow and can produce side effects including "keto" breath and constipation. If you're interested in the health benefits of ketosis, ask your provider if the diet may be right for you.

Ref From –

Blood-Brain Barrier (BBB): What It Is and Function. https://my.clevelandclinic.org/health/body/24931-blood-brain-barrier-bbb

https://www.ncbi.nlm.nih.gov/pmc/articles/PMC7699472/#:~:text=The%20uptake%20of%20ketone%20bodies,throughout%20the%20brain%20%5B9%5D.

https://my.clevelandclinic.org/health/articles/24003-ketosis

https://www.ncbi.nlm.nih.gov/books/NBK519556/#:~:text=The%20blood%2Dbrain%20barrier%20(BBB,the%20blood%20and%20the%20brain.

Chapter – 13

What are Micronutrients and Macronutrients

Nutritional science principally distinguishes two classes among its classifications: macronutrients and micronutrients. **Macronutrients can be considered as the main components of different tissues, and they constitute the total amount of caloric intake, meaning the principal energy source of the human body; they are mainly distinguished in carbohydrates, proteins, and lipids.**

Micronutrients are diet components that do not significantly contribute to caloric intake but can still be considered crucial for health and vital functions, even if needed in smaller amounts. They principally include vitamins (both fat-soluble and water-soluble) and minerals.

Since both macro and micronutrient deficiencies can have a significant impact on the whole function and physiology of the human body, particularly on the various processes of growth, a step-by-step distinction and knowledge regarding their influence during the different times of paediatric age can be a helpful guide for the clinical practice and constitutes the main propose of this review.

Micronutrients

The infant needs micronutrients to support his physiological growth. Breast milk is adequate for an infant's nutritional needs, including most micronutrients in the first semester of life: it is rich in minerals and vitamins. Breast milk is relatively deficient in iron (Fe) and zinc (Zn), but, while on the one hand, they are very efficiently absorbed, on the other, their deposits formed during pregnancy may support physiologic needs in the first 6 months of life, up to the beginning of weaning. If an exogenous source of iron is not provided during the second half of infancy, exclusively breastfed infants are at risk of becoming iron deficient. Among minerals, calcium (Ca) and phosphorus (P) are essential for growth, and breast milk is indeed rich in them. Iodine is necessary for the fetus and infant's growth; its early deficiency causes hypothyroidism and cretinism. Breast milk supplies adequate iodine content to the infant if the mother introduces a diet of about

250 µg/day of this mineral. The intake of artificially enriched food, such as salt, allows for the iodine requirement to be achieved.

After the sixth month of life, breast milk is not sufficient to meet nutritional needs, and the introduction of complementary foods should begin. The pediatrician should be able to advise parents about the more appropriate introduction of new foods.

Macronutrients

Lipid requirement

Lipids are the macronutrients with higher calorie density. They allow the absorption of fat-soluble vitamins and are a source of essential fatty acids, ARA, EPA, DHA, and cholesterol. The intake of lipids is necessary in the first two years to support brain growth and development. In early childhood, the quality of fats consumed is more important than quantity.

Carbohydrate requirements

Carbohydrates are macronutrients with an exclusively energetic function (4 Kcal/g). They are indispensable for the metabolism of erythrocytes and vital organs such as kidney and brain.

Micronutrients

During early childhood, minerals and vitamins are essential for growth. After weaning, a varied diet is necessary to obtain adequate micronutrient intake. Calcium, phosphorus, and vitamin D are crucial for bone growth; iodine allows thyroid hormone synthesis and brain myelination; iron is mainly necessary for synthesizing red blood cells and new tissues. Finally, zinc is essential for the growth and regulation of the immune system.

Calcium

Calcium needs depends on the development of bone mass. Ca is mainly contained in milk, cheese, yoghurt and vegetables. Calcium deficiency in pre-adolescent age causes rickets.

Phosphorus

Phosphorus is widely contained in foods, particularly in cereals, wholemeal flours, eggs, legumes, fish, milk, cheeses and meat. Given the wide distribution of phosphorus in food, deficiencies linked to insufficient food intake are rare. Chronically insufficient intake can compromise growth and cause rickets.

Iodine

Iodine is essential for thyroid hormones synthesis. Its deficiency in early childhood can cause goitre and hypothyroidism.

Vitamin D

Vitamin D is essential for calcium absorption and bone tissue synthesis. Its receptor is expressed in different cells, and for this reason vitamin D plays its role in many nonskeletal mechanisms. Several studies support a relationship between serum 25-hydroxyvitamin D [25-(OH)D] levels and chronic metabolic, cardiovascular, and neoplastic diseases.

The skin can produce vitamin D after solar exposure, but the child can also introduce it with food. Cod liver oil, fish, pork liver, eggs and butter naturally contain D vitamin. Vitamin D deficiency in early childhood causes rickets.

Vitamin A

Vitamin A is contained in the liver of several animal species, as well as in eggs and fish. Vegetables contain carotenoids, a pro vitamin from which vitamin A derives. Retinol equivalent (RE) is the unit of measure used to describe the total quantity of vitamin A (vitamin and pro vitamin) in the diet. In developing countries, at high risk of malnutrition, vitamin A deficiency can cause night blindness. The lack of vitamin A, in some cases, causes conjunctival lesions, xerophthalmia or keratomalacia

Zinc

Zn is essential for growth, immune and gastrointestinal systems. Infant and toddler have an increased risk of Zn deficiency, a condition characterized by growth retardation, alopecia, diarrhea, skin lesions, loss of appetite and reduced antioxidant defenses. Another effect of Zn deficiency is impaired immune function.

Iron

The infant faces the first months of life with iron deposits acquired in utero, but after the sixth month dietary intake becomes essential. Meat, fish, cereals and eggs contain iron. Among tissues, bone marrow requires large quantities of iron for the synthesis of hemoglobin. Iron deficiency manifests itself with reduced physical performance due to the reduction of hemoglobin and myoglobin. Other consequences of the deficiency state are the impairment of the immune response with a reduction in the function of macrophages and neutrophils and a decrease of T lymphocytes. Brain iron deficiency causes reduced myelin and neurotransmitter synthesis with impaired movement, memory and perception control. Infants iron deficiency can cause irreversible brain damage.

Vitamin K

Vitamin K activates coagulation factors. Its severe deficiency causes bleeding. From weaning, the infant can acquire this vitamin through the diet. The foods that contain it most are green leafy vegetables, such as broccoli, lettuce and spinach.

Vitamin C (ascorbic acid)

Vitamin C participates in enzymatic reactions such as biosynthesis of collagen, carnitine and norepinephrine. It also works in non-enzymatic reactions, such as neutralization of reactive oxygen species. Vitamin C is found mainly in vegetables. Its deficiency is usually due to an inadequate intake and causes scurvy. This disease manifests itself with fatigue, weight loss, arthralgias and gums bleedings; subsequently, if left untreated, it induces the formation of hematomas (especially in the lower limbs). It also causes impaired scarring, tooth loss, bruising and bleeding in many organs, sometimes with a fatal outcome.

B vitamins

B vitamins are water-soluble nutrients involved in different metabolic processes, allowing the use of carbohydrates, lipids and proteins to obtain energy. They are fundamental for the development of all organs and systems, particularly the nervous system. Humans cannot synthesize B vitamins and need to introduce these nutrients with food. The infant takes these vitamins from breast milk, subsequently, from the moment of weaning, the intake of the B complex vitamins depends on other foods. A diet rich in whole grains, green leafy vegetables, oilseeds, dried fruit and legumes ensures a sufficient dose of B vitamins.

Thiamine (B1)

Thiamine acts as a coenzyme in the metabolism of carbohydrates, branched-chain amino acids, fatty acids and nucleotides. It allows the synthesis of acetylcholine and also act as a neuromodulator. For these reasons, it may be qualified as essential for the central and peripheral nervous system. A severe and protracted thiamine deficiency causes beriberi, a disease characterized by neurological, cardiovascular and gastroenteric manifestations.

Riboflavin (B2)

Riboflavin participates in oxidation-reduction reactions in metabolic pathways involving carbohydrates, lipids and amino acids. Riboflavin deficiency causes skin lesions, anemia, neuropathy and corneal vascularization.

Niacin (B3)

Niacin is the constituent of enzymatic cofactors that participate in many metabolic reactions of oxidation-reduction. It contributes to the metabolism of fats, carbohydrates and proteins. It helps the nervous system in its functions, improves circulation and reduces blood cholesterol. Severe niacin deficiency leads to pellagra, a disease characterized by dermatitis, diarrhoea and dementia.

Pyridoxine (B6)

Pyridoxine is a cofactor in the folate cycle, and the synthesis of neurotransmitters such as dopamine, serotonin, GABA, norepinephrine and melatonin. B6 deficiency can lead to reduced production of these neurotransmitters and, consequently, to sleep and behaviour disorders, cognitive decline, peripheral neuritis, cardiovascular diseases, anaemia and dermatitis.

Biotin (B8)

Biotin is the coenzyme of some carboxylases involved in energy metabolism, in the synthesis of fatty acids and the catabolism of amino acids. A primary biotin deficiency is rare. Biotin deficiency can occur in cases of protracted parenteral nutrition and can manifest itself with eczema, mood alteration and immunodeficiency.

Folic acid (B9)

Folic acid is necessary for DNA synthesis, repair and methylation reactions. Therefore, it allows cell division reactions in case of rapid growth and also enables the metabolism of homocysteine. Humans cannot produce folic acid, so they find this vitamin in foods such as green leafy vegetables, legumes, cereals, fruit and liver. Folic acid deficiency causes anaemia and an increase in serum homocysteine.

Vitamin B12

Vitamin B12 participates in the metabolism of homocysteine and allows the synthesis of DNA and haemoglobin.

Food of animal origin, such as sheep liver, veal, chicken and turkey are rich in vitamin B12. Some fish also contain vitamin B12. Eggs and cheeses contain B12 in much lower concentrations than meat and fish. B12 deficiency causes megaloblastic anaemia, increased serum homocysteine and can lead to permanent neurological damage.

Think of macronutrients as the lead actors in the production of your body and micronutrients as the supporting cast. Each is vital to a successful performance.

Registered dietitian Julia Zumpano, RD, LD, explains exactly what macronutrients and micronutrients are and why they're so important to your health.

As the main nutrients found in food, macronutrients maintain your body's structure and functioning. You typically need a large amount of macronutrients to keep your body working properly. But don't stress: macronutrients come from proteins, fats and carbohydrates, which give your body energy in the form of calories.

Examples of macronutrients

During digestion, foods that tend to fall into one of the three macronutrients are broken down to be used for different functions.

Macronutrients include:

Carbohydrates: As the main source of energy, carbs break down into glucose and aid digestion and fullness. Carbs include bread, rice, pasta, grains, fruits, starchy vegetables, beans, milk and yogurt. They provide 4 calories per gram.

Fat: Fats are broken down into fatty acids and glycerol and provide fat-soluble vitamins A, D, E and K. Foods like nuts, seeds, oils, butter, sour cream, mayo and cream cheese provide 9 calories per gram.

Protein: Protein helps build and repair muscle, tissues and organs, as well as aid in hormone regulation. Foods like meat, poultry, fish, eggs, cheese, cottage cheese, plain Greek yogurt and tofu provide 4 calories per gram.

Minerals that are good examples of micronutrients include:

Calcium: This mineral helps build strong bones and teeth and helps with muscle function. Foods include yogurt, orange juice, cheese and milk.

Magnesium: Found in foods like pumpkin seeds, almonds and spinach, this mineral aids in the regulation of blood pressure.

Sodium: You need sodium for optimal fluid balance and to maintain your blood pressure.

Potassium: Potassium helps with muscle function and nerve transmission. You can find potassium in foods like apricots, lentils, prunes and raisins.

Ref From –

Macronutrient balance and micronutrient amounts through growth and development | Italian Journal of Pediatrics | Full Text. https://ijponline.biomedcentral.com/articles/10.1186/s13052-021-01061-0

https://www.ncbi.nlm.nih.gov/pmc/articles/PMC10002864/

https://health.clevelandclinic.org/macronutrients-vs-micronutrients

Chapter – 14

Glutathione

Glutathione is a tripeptide composed of glutamate (glutamic acid), cysteine, and glycine residues with an unusual peptide bond between the α-amine of cysteine and the side-chain carboxylate of glutamate.

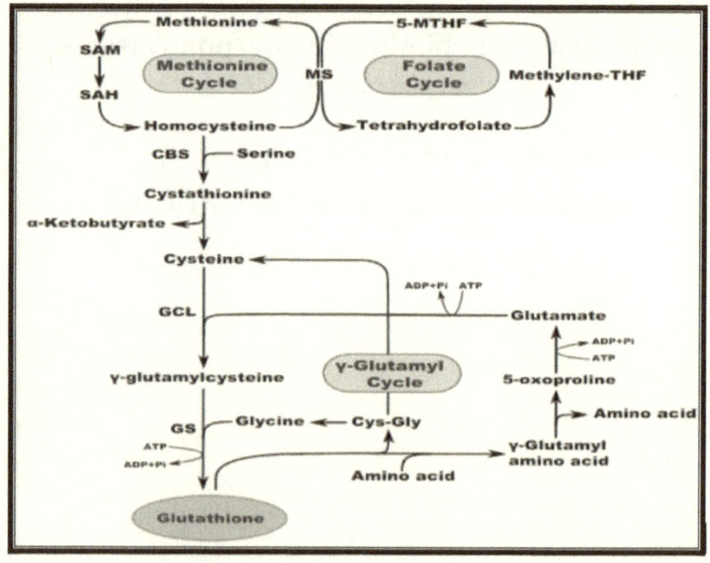

Picture Courtesy From - https://www.mdpi.com/1420-3049/27/1/324

Glutathione (master antioxidant) boosts the utilization and recycling of other antioxidants, namely vitamins C and E, alpha-lipoic acid, and CoQ10. Glutathione (GSH), the major intracellular thiol compound, is a ubiquitous tripeptide produced by most mammalian cells, and it is the main antioxidant defense mechanism against reactive oxygen species (ROS) and electrophiles. GSH (γ-glutamyl-cysteinyl-glycine) is synthesized de novo in two sequential enzymatic ATP-dependent reactions.

Glutathione is one of the most abundant endogenous antioxidants found in the body and plays an important role in maintaining exogenous antioxidants (i.e., vitamins C and E) in their active reduced roles. It is produced in the liver and synthesized from cysteine, glutamic acid, and glycine. Glutathione also functions as a detoxification agent of carcinogens and harmful foreign compounds. Glutathione supplementation in infertile men has been demonstrated to improve sperm motility.

Glutathione (GSH) is an essential antioxidant molecule that prevents damage caused by reactive oxygen species, such as free radicals and peroxides. It can, therefore, regulate the cell response to oxidative stress.

Glutathione (GSH) is often referred to as the "mother of all antioxidants." It is an important antioxidant that is conventionally found in bacteria, animals, fungi, and plants. It is highly abundant in all of the compartments of an organism's cells. Moreover, it is the major soluble antioxidant, and its ratio helps to determine oxidative stress. Glutathione detoxifies the lipid peroxides and hydrogen peroxide through the action of GSH-Px.

Glutathione (GSH) is also regarded as an intracellular thiol that eradicates the free radicals and causes a reduction in the levels of hydrogen peroxide.

There are two different forms of glutathione: reduced glutathione (GSH, or L-glutathione), which is the active form, and oxidized glutathione (GSSG), which is the inactive state. As GSH patrols the cellular environment and puts out oxidative "free radical" fires, they become oxidized and inactive, thus turning into GSSG.

Fortunately, the glutathione reductase enzyme can recycle inactive GSSG back into the active GSH form. When this enzyme is overwhelmed, and too much oxidized GSSG accumulates (compared to the active GSH), cells become susceptible to damage.

Mitochondria are the "power plants" of each cell, converting food into ATP (adenosine triphosphate) for all of our cells' energy needs.

Mitochondria do much more than pump out energy. They also have their DNA; they can communicate information, sense danger when the cell energy levels drop, and are even involved in sending the final "death" message (apoptosis) when a cell is damaged beyond repair and needs to die.

Glutathione has many vital roles in your health and well-being. Four of the most critical are:

- Aging defense

- Antioxidant protection

- Detoxification

- Energy production

Glutathione is also responsible for:

- Cysteine carrier/storage

- Cell signaling

- Enzyme function

- Gene expression

- Cell differentiation/proliferation

The more glutathione in your body, the healthier your cells, and mitochondria. The less glutathione in your body, the more likely you are to have a cellular breakdown, increased risk of disease, and cellular death.

Methylation

Methylation is critical for human survival. For example, it is like an electrical switch that turns genes on and off. Methylation is also integral to how we function every second of the day. It regulates neurotransmitters, brain function, mood, energy, and hormone levels. It is fair to say that methylation is almost synonymous with physical function.

Homocysteine is one of the most well-studied products of the methylation cycle, which is the common link between methylation and making glutathione, also called the "trans-sulfuration" pathway. (See figure below showing the methylation and transsulfuration pathways.)

Glutathione production starts with the amino acid cysteine. This first step is also the most critical "rate-limiting" step. As noted above, the usual cysteine source comes from homocysteine, a significant product of the methylation cycle. So, making glutathione depends on a well-functioning methylation cycle that provides enough homocysteine.

Conversely, suppose the glutathione production process (or the "trans-sulfuration" pathway) is malfunctioning. In that case, the process backs up, and homocysteine levels accumulate, putting additional strain on the methylation cycle to remove it. High homocysteine levels are problematic because they have been linked to heart disease and atherosclerosis. In people deficient or with mutations in the enzymes that catalyze the production of glutathione from homocysteine, the methylation cycle will be under pressure to remove excess homocysteine.

One such enzyme is cystathionine beta-synthase (CBS), which catalyzes the first and most important (rate-limiting) step in trans-sulfuration from homocysteine to cystathionine. Individuals with CBS mutations will be slow to make glutathione.

Flipping this around, individuals who have "slow" methylation cycle enzymes will have lower homocysteine levels. Slow methylation can directly affect and lower glutathione levels since it's the first step in making glutathione.

By now, you may have heard of the most famous enzymes—MTHFR and MTR— regulating the speed of the methylation cycle as physicians are directly now ordering more and more genetic testing. These enzymes control the methylation cycle speed and efficiency, determine homocysteine levels, and indirectly affect glutathione production. In conclusion, for those of you who have been tested and know you have MTHFR and MTRR or CBS mutations, you might be struggling with low glutathione production and levels without realizing it. Methylation is a critical process—as well as a complicated one.

The key to remember is that low methylation equals low glutathione and that low glutathione slows methylation. They are interdependent. The solution? Maintain normal glutathione levels, and all will be good.

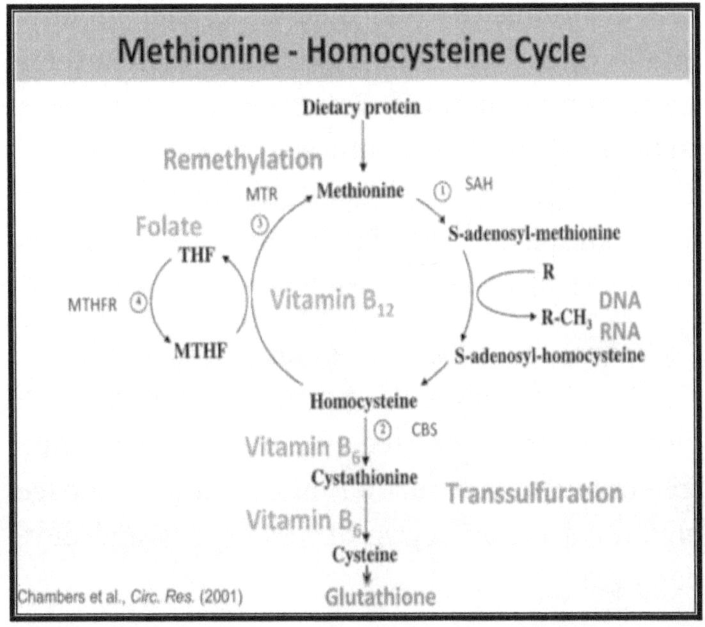

Picture Courtesy From - https://www.researchgate.
net/figure/The-methionine-homocysteine-cycle-
contains-re-methylation-and-transsulfuration_
fig6_286447602

Antioxidants are the "anti-agers" of the nutrient world, working to protect your body from free radical or "oxidative" damage. Every time you eat, breathe, or move, your body uses fuel created from the food you eat to produce energy. But just as a car using gas to make energy releases harmful byproducts of this process as exhaust, so does your own body's energy-producing efforts have a dangerous byproduct—free radicals.

Energy production is located inside mitochondria in all cells (except red blood cells). Glutathione protects mitochondria from free radicals or other "oxidative" damage. If mitochondria are attacked and damaged by oxidative molecules, they slow down and make less ATP. With less ATP, the rest of the cell also becomes sluggish.

To make things worse, damaged mitochondria also become more error-prone and start to create more "exhaust" or free radicals. In turn, these free radicals cause further mitochondrial damage, creating a vicious cycle of less energy and more damage.

Stress also affects energy production. The higher the energy needs (higher metabolism, exercise, stress, etc.), the harder the mitochondria have to work and the more free radicals they produce.

GSH binds these free radicals and relieves "oxidative stress" on the mitochondria and the rest of the cell. In doing so, GSH becomes oxidized and converts to GSSG.

With the help of the enzyme glutathione reductase, it can be recycled and turned back into active glutathione or GSH. However, if this process is overwhelmed or doesn't work

correctly, GSSG accumulates, and the ratio of GSH/GSSG becomes distorted.

The ratio of GSH/GSSG can be measured and is a very reliable measure of "oxidative stress" or how fast we are aging and deteriorating. This means we can measure how susceptible our cell's DNA, cell membranes, proteins, and cholesterol are to damage.

These symptoms are not only associated with many chronic diseases. Still, they are also a result of **"mitochondrial dysfunction,"** which occurs when mitochondria lose the protection of GSH, free radicals attack the mitochondria and cellular energy decreases.

Autoimmune conditions like multiple sclerosis, Crohn's disease, rheumatoid arthritis, diabetes, and more all have "mitochondrial dysfunction," low levels of GSH, and profound fatigue.

Glutathione deficiency makes you vulnerable to oxidative stress and inflammation, both of which are markers of accelerated aging and chronic illness. If you have too little GSH, you can't fight off your cell's mitochondria. As a result, you start to feel more tired because the mitochondria are less efficient when they get oxidized or "rusted."

The free radical damage caused by oxidation then triggers your immune system to clean up the damage, which results in inflammation.

It's no surprise that depleted levels of glutathione can increase your risk for several adverse health conditions, including heart disease and diabetes, among others.

This is even more problematic, given the number of factors that can deplete glutathione levels.

In addition to natural aging, environmental causes include:

- Chronic exposure to chemical toxins

- Alcohol use

- Smoking

- Pollution

- Poor diet

- Stress

- UV radiation exposure

Certain illnesses are known to decrease glutathione levels. Some of the more common low glutathione-related diseases are:

- Macular degeneration

- Parkinson's disease

- Diabetes

- Hepatitis

- Cancer

- COPD

- Alzheimer's disease

- Liver disease

- Stroke

- Heart disease

- Infertility

Brain Health

As we age, it's not uncommon to experience a bit of forgetfulness or maybe have difficulty concentrating or remembering names or where we left our car keys. This is technically called "neuro-degeneration," a process by which the neurons in our brains become damaged and may even die, leaving us with "shrinking" brains that don't function to their full capacity. While this process is unavoidable as we age, it can be slow or even reversed, and glutathione (GSH) plays an important role.

Certain brain disorders have accelerated neuro-degeneration that give us clues. For example, Parkinson's and Alzheimer's diseases have high levels of oxidative stress and damage to the brain with correspondingly low active glutathione (GSH) levels. GSH can help ease and decrease the rate of damage to neural tissue.

Heart and Cardiovascular System

Glutathione prevents heart attack and stroke by neutralizing the "lipid oxidation" process. This is important because virtually all heart disease starts with accumulating arterial plaques or deposits inside the arteries' walls.

Coronary and arterial plaque (atherosclerosis) develops gradually as cholesterol particles such as LDL in the blood are "lipid oxidized" and damage the blood vessels' lining, forming a plaque. When these plaques eventually rupture and break off, they cause clogs that block blood flow and cause heart attacks or strokes.

With the help of an enzyme called glutathione peroxidase, glutathione inactivates the superoxide, free radicals, hydrogen peroxide, lipid peroxides, and peroxynitrite that cause this "lipid oxidation" and wreak havoc on your health. In this way, glutathione helps to prevent damage and lowers the risk of heart attacks.

Inflammation

Inflammation is present in virtually every chronic illness, from diabetes and heart disease to cancer. Any injury can incite an inflammatory response, whether you are talking about trauma, an infection, toxins, or allergies.

When an injury is detected, your body produces an enzyme called cyclooxygenase-2 (or COX-2), which sets this inflammation process into motion.

In turn, COX-2 signals to produce a short-lived signaling molecule called series-2 prostaglandins. These pro-inflammatory hormones encourage inflammation and help the body heal the injured area.

Once your body has done its job, it needs to restore your body to normal and switch off these hormones. To do that, it releases COX-1 enzymes that signal for the release of series-1 and series-3 prostaglandins, which are anti-inflammatory.

In the real world, environmental toxins, diet, stress, and other lifestyle issues have disabled this system's checks and balances, encouraging your body to make more pro-inflammatory prostaglandins and less anti-inflammatory ones. As a result, many people suffer from chronic, systemic inflammation. It appears that glutathione (GSH) controls when inflammation increases or decreases as needed by instructing and influencing our immune white cells. Additionally, autoimmune disease appears to be hallmarked by imbalanced glutathione levels.

Skin Health

When it comes to skin problems like acne, wrinkles, dryness, eczema, or puffy eyes, exposure to UV sun, wind, outdoor activities with inadequate nutrition, stress, lack of exercise, and hormonal changes can take its toll on exposed skin, resulting in dry, wrinkled skin and age spots.

Thanks to glutathione, you can solve the problem internally and have cells heal and regenerate themselves.

Glutathione not only decreases the melanin (pigmentation) of skin but has also been found to reduce the appearance of wrinkles and increase the skin's elasticity.

Glutathione mainly affects skin pigment production by inhibiting tyrosinase, one of the

enzymes that makes melanin. Interestingly, in one study, both GSH and GSSG achieved the skin-lightening effect.

Ref From –

https://www.sciencedirect.com/topics/medicine-and-dentistry/glutathione

https://coremedscience.com/blogs/wellness/glutathione-the-master-antioxidant

https://www.frontiersin.org/journals/pharmacology/articles/10.3389/fphar.2014.00151/full

Chapter – 15

NAD

Mitochondria are producing a low level of free radicals all of the time. Since mitochondrial DNA is stored within the mitochondria, it is particularly exposed to this source of damage, and is thought to mutate roughly 10 times faster than nuclear DNA. Cellular levels of antioxidants and free radical-scavengers that protect against this damage also decline with age. Consequently, an increasing proportion of the mitochondria within our cells become dysfunctional as we age, meaning that they do not produce ATP as efficiently as they should. These dysfunctional mitochondria are supposed to be destroyed by the cell in a recycling process called autophagy (also called mitophagy where mitochondria specifically are concerned). However, this process becomes disrupted with age (see loss of proteostasis). Consequently, entire cells can eventually be 'taken over' by dysfunctional mitochondria.

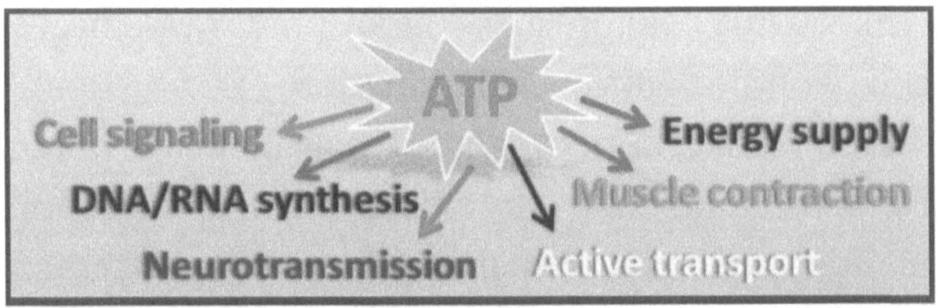

Pic Courtesy from - https://www.youtube.com/watch?v=VP8WnuK7PJU

One way in which mitochondrial dysfunction may contribute to ageing and its diseases is through inefficient ATP production. ATP is an essential fuel for cells of all types. It provides the energy necessary for muscles to contract, for neurons to send electrical impulses, and for cells to control the concentration of ions inside their membranes. It is suggested that a deficit of energy may be responsible for some aspects of ageing, particularly in tissues that demand a lot of energy like muscle. ATP is also required for processes that remove toxic proteins such as amyloid, thought to play a role in many age-associated diseases including Alzheimer's.

Mitochondria are important to the cell in more ways than merely producing energy. Products of mitochondrial metabolism act as signaling molecules that influence a wide range of cellular processes, including apoptosis (cell suicide), autophagy (the disposal of damaged cell components), inflammation, and glucose metabolism. These processes may all be disrupted by mitochondrial dysfunction. The accumulation of dysfunctional mitochondria also depletes an essential molecule called nicotinamide adenine dinucleotide (NAD). NAD is necessary for the production of ATP, but is also used in many processes that are vital to the cell such as the repair of damaged DNA, the recycling of damaged cell components, and the growth and division of existing mitochondria.

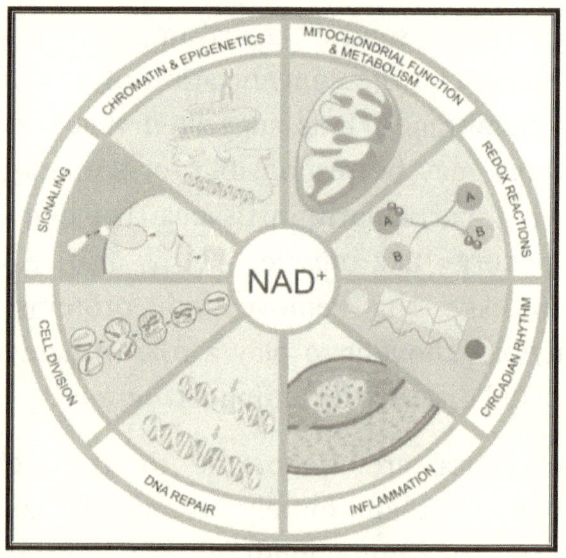

Pic Courtesy - https://www.sciencedirect.com/ science/article/pii/S1550413118301220#fig2

How Can Mitochondrial Dysfunction Be Measured?

Various approaches can be used to measure mitochondrial function in isolated mitochondria and cells. Simply studying the microscopic appearance of mitochondria can indicate mitochondrial function, as can the expression of specific genes involved in mitochondrial growth and division. The rate of energy production itself can be studied by measuring the electrical potential across the mitochondrial membrane, which is necessary to power the proteins that are used to produce ATP. Levels of oxidative stress, the result of free radical production, can also measure mitochondrial function. However, culturing cells in a dish changes the behavior of their mitochondria, as they are not exposed to the same signals and do not metabolize the same nutrients as when they are within a living organism. Measuring mitochondrial dysfunction within living organisms may be more relevant to human health, though it is also more challenging and less precise.

Changes in patterns of gene expression and protein production may suggest altered mitochondrial function. Since most of the oxygen we breathe is consumed within mitochondria to produce ATP, measuring the rate of oxygen consumption or speed of recovery from exercise may be used as an indicator of mitochondrial function. However, this relationship is not straightforward since ATP production may be influenced by other factors, such as the supply of nutrients to the mitochondria. Imaging techniques such as positron emission tomography (PET) allow the uptake of nutrients by cells to be measured directly, providing a dynamic picture of energy use in different locations. Nutrient uptake is indicative of mitochondrial energy production, but again, the relationship is not straightforward.

How Might Mitochondrial Dysfunction Be Fixed?

One of the leading underlying causes of mitochondrial dysfunction is mutations in the mtDNA. As discussed in the first article of this series, such mutations are challenging to reverse, but this does not mean that mitochondrial dysfunction could not be prevented or even reversed through other means. For example, drugs could be used to protect mitochondria against damage from free radicals and promote the destruction and replacement of damaged mitochondria. One compound with therapeutic potential is **coenzyme Q10**, an enzyme that can protect against mitochondrial damage, improve ATP production and promote growth of new mitochondria.

There is also interest in therapies targeting NAD levels and the enzymes that depend on NAD, namely sirtuins. In response to NAD levels, sirtuins regulate the cell's metabolism in response to nutrient availability and the expression of genes in response to damage. Compounds that boost NAD (such as vitamin B3) or activate sirtuins (such as resveratrol and pterostilbene) appear to have anti-aging properties in animals. However, human data remains sparse, and the health benefits of boosting NAD and sirtuin activity in humans remain controversial.

One of the most ambitious methods for tackling mitochondrial dysfunction is to insert backup copies of mitochondrial DNA into the cell nuclei, which is a much safer environment than the interior of the mitochondria.

This is not an unrealistic goal since, during our shared evolution, almost all of the thousand or so mitochondrial genes have already been driven out of the mitochondria and into the cell nucleus by evolutionary pressure. The SENS research foundation's mitoSENS project has already succeeded in making backups of two mitochondrial genes function correctly in human cell nuclei. The next step is to study the effects of these backups in mouse models.

Ref From –

https://www.gowinglife.com/what-is-mitochondrial-dysfunction-the-hallmarks-of-ageing-series/

The Hallmarks of Aging:

https://dx.doi.org/10.1016%2Fj.cell.2013.05.039

Assessing mitochondrial dysfunction in cells:
https://dx.doi.org/10.1042%2FBJ20110162

Mitochondrial Dysfunction in Aging
and Diseases of Aging: https://dx.doi.
org/10.3390%2Fbiology8020048

Skeletal muscle aging and the mitochondria:
https://dx.doi.org/10.1016%2Fj.tem.2012.12.003

Role of Mitochondria in the Mechanism(s) of
Action of Metformin: https://doi.org/10.3389/
fendo.2019.00294

Therapeutic potential of coenzyme Q10 in
mitochondrial dysfunction during tacrolimus-

induced beta cell injury: https://doi.
org/10.1038/s41598-019-44475-x

Slowing ageing by design: the rise of NAD+ and
sirtuin-activating compounds: https://dx.doi.
org/10.1038%2Fnrm.2016.93

Chapter – 16

What is NMN

Is there a Difference Between NAD+, NMN, and NR Supplements?

NMN Supplements

NAD, or nicotinamide adenine dinucleotide, is a molecule or coenzyme found in every cell in your body. The part of the NAD coenzyme that performs essential functions is known as NAD+.

NAD+ has two important roles—helping proteins regulate the functions in your cells and turning nutrients into energy. The process of converting food to energy is what helps you stay alive. NAD+ supports all the crucial functions of your cells, affecting your sleeping and waking cycles, your neurological health, and the health of your vital organs.

As people grow older, their levels of NAD+ naturally decrease. Stress on the body and mind may cause NAD+ levels to decline even further. Low NAD+ levels increase your risk for many health conditions, like heart disease, Alzheimer's disease, and Type 2 diabetes.

NAD+ and Nicotinamide mononucleotide (NMN) are in the same cellular pathway; NMN is the precursor of NAD+, and the body turns naturally into NAD+. NMN is more stable than NAD+. Boosting NMN levels can increase NAD+ biosynthesis and has been shown to mitigate aging-related dysfunctions. A recent Japanese double-blind, randomized trial on healthy humans shows that NMN is safe and well-tolerated and can replenish cellular NAD+ levels to mitigate aging-related NAD+ biosynthesis.

Nicotinamide riboside (NR) Supplements

NR is another intermediate in NAD+ biosynthesis, and increasing NR can also increase NAD+ levels. A combination of NR and pterostilbene significantly increases the concentration of NAD+ in a dose-dependent manner in a randomized, double-blind, and placebo-controlled study in a population of 120 healthy adults between the ages of 60 and 80 years.

NAD+ (Nicotinamide Adenine Dinucleotide): A crucial molecule in the body, it plays a role in cellular energy production, DNA repair, and other vital processes. Direct supplementation with NAD+ might not be common due to potential issues with cellular uptake.

NMN (Nicotinamide Mononucleotide): It is a direct precursor to NAD+, meaning the body can convert it to NAD+. Some research suggests NMN might have potential anti-aging benefits, though human studies are ongoing.

NMN, or nicotinamide mononucleotide, is a cellular molecule that's considered a precursor to NAD+. This means NMN will chemically transform into NAD+ in your cells. When you think about NMN vs. NAD+, remember that these two molecules work together to metabolize energy in your cells.

After NMN is absorbed into the bloodstream, it converts into NAD+. Then, the NAD+ is stored in the muscles and vital organs, helping them function at peak condition.

One of the fastest ways to stimulate your body's NAD+ production is to increase levels of NMN. There is evidence that NMN's boost to NAD levels may be able to improve health in the heart, liver, muscles, and kidneys and even slow down the progression of diabetes and Alzheimer's.

Beyond their role in making NAD+, NMN molecules have their own health benefits. NMN supplements can aid your body's insulin production and glucose tolerance. They have positive effects on overall metabolism as well.

NMN supplements are the easiest way to raise NMN levels. Low volumes of NMN are also found in certain foods like broccoli, avocado, cucumber, and cabbage.

NR (Nicotinamide Riboside): Another precursor to NAD+ that might be more readily converted into NAD+ in the body than NMN. Like NMN, research into NR's potential benefits is ongoing.

Each anti-aging molecule has its benefits and disadvantages, as shown in Table

NAD+	NMN	NR
Benefits: Taking NAD+ directly can rapidly increase cellular NAD+. NAD+ intravenous applications mostly achieve this. Studies have shown that NAD+ can boost cellular energy levels and activate cell anti-aging genes. It also enhances brain functions. Supports muscle endurance and strength.	**Benefits**: Taking NMN is the best way to boost NAD+. The NMN is more stable than NAD+ and stably increases NAD+ pools for a long. Increasing NMN has all the benefits of NAD+ and specifically helps with vision and hearing loss related to aging. Activates longevity and anti-aging promoting sirtuin genes.	**Benefits**: Taking NR can also increase NAD+. A double-blind, randomized trial on 30 healthy individuals showed that nicotinamide riboside effectively elevated NAD+ levels in humans and tolerated well.
Best Delivery: Intravenous and sublingual	**Best Delivery**: Sublingual and transdermal. NMN also has relatively better bioavailability for gut-based delivery in capsule forms.	**Best Delivery**: NR also has relatively better bio-availability for gut-based delivery in capsule forms.
Recommended best for Athletic performance and increased energy metabolism for energy-deficient vulnerable groups.	**Recommended best for** Antiaging purposes and increasing mitochondrial functions.	**Recommended best for** Antiaging purposes and increasing mitochondrial functions.

Brain Health Benefits: NAD+ & NMN Supplements

Improving brain health functions stands out among the most important age-related benefits. As a major energy-churning organ, NAD+ depletion in brain cells due to aging leads to several issues, such as memory and cognitive impairment and brain energy imbalance, which are also related to vascular aging.

One of the major issues with drug addiction patients is low brain energy levels, resulting in mood alterations, distorted time and sensory perception, decreased memory, and problem-solving. NAD+ has been clinically used since the 1960s to help break free from chemical dependence. NAD+ is an all-natural vitamin B derivative with minimal side effects. Similarly, increasing NAD+ using NMN may significantly benefit neurodegenerative diseases such as Alzheimer's and Parkinson's.

Innovations in NAD and NMN Supplements

Dr. David Sinclair at Harvard University is the pioneer in elucidating the prominent functions of NAD+ and NMN in the anti-aging part, especially the sirtuin activation pathways.

Several groups are working on NAD and NMN clinical programs and better delivery methods. A new study has shown that boosting NMN and NAD+ in late life can restore and rescue female reproductive function in mammals, suggesting the future of NMN supplements in reproductive health.

How To Naturally Maintain Healthy NAD+ Levels

Besides supplements, a balanced diet and active lifestyle can help sustain good cellular health. A combination of vegetarian and fiber-rich food can increase the vital gut microbiome metabolism, maintaining a healthy cellular pool of NAD+. An inactive lifestyle and unhealthy diets with saturated fatty acid and sugar can significantly reduce the NAD+ and rapidly prompt the cell to age. A healthy regimen of NAD/NMN supplements with a healthy diet may easily reverse aging by 10-15 years.

Ref From –

https://biomprobiotics.com/the-ultimate-guide-to-understanding-nmn-and-nad-supplements/#:~:text=NAD%2B%20and%20Nicotinamide%20mononucleotide%20

(NMN,to%20mitigate%20aging%2Drelated%20 dysfunctions.

https://www.springfieldwellnesscenter.com/ nad-blog/nad-vs-nmn-whats-the-difference/

Chapter – 17

Nitric Oxide

Nitric oxide is a gaseous signaling molecule that modulates physiological functions in the human body. It is a neurotransmitter that affects blood flow, muscle contractility, mitochondrial respiration, glucose homeostasis, and immune function.

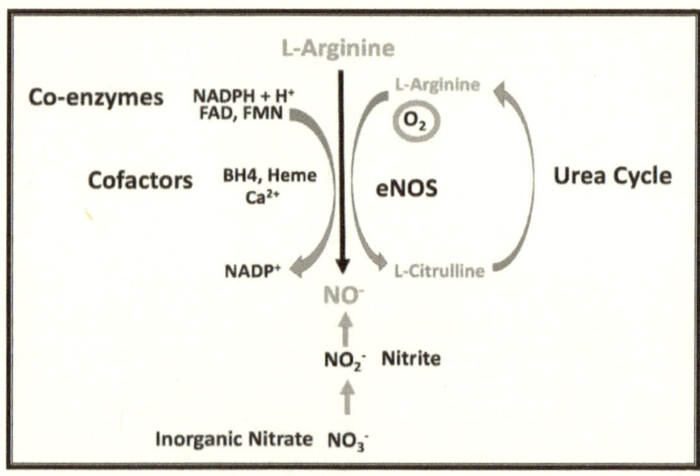

Picture Courtesy From - https://humann.com/blogs/ explore/how-the-body-makes-nitric-oxide

Nitric oxide can be produced both in the presence and absence of oxygen. Additionally, oral bacteria residing on the tongue's surface can reduce both endogenous and exogenous NO3– to NO2– via the intestinal-gastric circulation.

Nitric oxide is generated by oxidation of the amino acid l-arginine in a reaction catalyzed by nitric oxide synthase (NOS). Nitric oxide oxidation produces nitrite and nitrate.

Nitric oxide (NO) is an essential component of the human body, involved in blood vessel dilation, hormone release stimulation, signaling and neurotransmission regulation. Nitric oxide is synthesized by nitric-oxide-synthase-dependent and --independent pathways. Nitric oxide supplementation improves cardiac health, enhances performance during exercise, reduces high blood pressure during pregnancy, reduces erectile dysfunction and improves healing processes and respiratory response. Nitric-oxide-associated benefits are mostly apparent in untrained or moderately trained individuals.

L-arginine and L-citrulline supplementation contributes to nitric oxide levels because L-arginine is directly involved in NO synthesis, whereas L-citrulline acts as an L-arginine precursor that is further converted to NO by a reaction catalyzed by NO synthase.

Nitric oxide or nitrogen monoxide (NO) is a component of the human body that may cause blood vessel dilation and the release of hormones such as human growth hormone (GH) and insulin.

L-arginine is a metabolically versatile amino acid that builds proteins and synthesizes NO. It is found naturally in dairy products, red meat, fish and poultry. Commercial manufacturers produce it as a pill, cream or powder. The human body converts consumed L-arginine into NO for different body functions.

L-citrulline is a non-essential amino acid found naturally in nuts, meat, watermelon, and legumes. It is manufactured commercially as a powder or pill. In the body, L-citrulline is synthesized by conversion of L-arginine to NO in a NOS-catalyzed reaction.

Considerations and Potential Side Effects

Nitric oxide supplementation is considered safe. However, there are a few side effects and contraindications to watch out for.

- Those taking blood pressure medication should avoid taking nitric oxide supplements as both decrease blood pressure, leading to hypotension.

- Side effects of L-arginine and L-citrulline include diarrhea, nausea, and vomiting.

It is important to consult with a functional medicine practitioner before supplementation to ensure it is best for you.

Ref from –

https://www.ncbi.nlm.nih.gov/pmc/articles/ PMC9710401/

https://www.sciencedirect.com/ topics/medicine-and-dentistry/nitric- oxide#:~:text=Nitric%20oxide%20is%20a%20 gaseous,presence%20and%20absence%20of%20 oxygen.

NMN vs NAD: Which is Better? – NutrBank. https://nutrbank.com/blogs/news/nmn-vs-nad- which-is-better

Student wins awards at science fair for college- assisted project » The Veterinary Page - College of Veterinary Medicine - University of Florida. https://veterinarypage.vetmed.ufl.edu/2017/05/24/student-wins-awards-at-science-fair-for-college-assisted-project/

Signal Transduction Questions and Answers - Sanfoundry. https://www.sanfoundry.com/cell-biology-questions-answers-role-no/

L-Citrulline. https://www.5thnutrisupply.com/sports-nutrition/l-citrulline

Bonetti, G., Kiani, A. K., Medori, M. C., Caruso, P., Manganotti, P., Fioretti, F., Nodari, S., Connelly, S. T., & Bertelli, M. (2022). Dietary supplements for improving nitric-oxide synthesis. https://doi.org/10.15167/2421-4248/jpmh2022.63.2S3.2766

https://www.rupahealth.com/post/the-anti-aging-effects-of-nitric-oxide-on-our-skin#:~:text=Nitric%20oxide%20is%20a%20promising,oxide%20can%20reverse%20premature%20aging

Chapter – 18

COQ10 and Anti-Aging

Mitochondrial diseases are not contagious and are not caused by anything a person does. They're caused by mutations, or changes, in genes — the cells' blueprints for making proteins.

Genes are responsible for building our bodies and are passed from parents to children, along with any mutations or defects they have. That means that mitochondrial diseases are inheritable, although they often affect members of the same family differently.

The genes involved in mitochondrial disease typically make proteins that work in or on the mitochondria. Within each mitochondrion (the singular form of mitochondria), these proteins make up part of an assembly line that uses fuel molecules derived from food to manufacture the energy molecule adenosine triphosphate (ATP) through oxidative phosphorylation. This highly efficient manufacturing process requires oxygen; outside the mitochondrion, there are less efficient ways of producing ATP without oxygen.

Proteins at the beginning of the mitochondrial assembly line act like cargo handlers, importing the fuel molecules — sugars and fats — into the mitochondrion. Next, other proteins break down the sugars and fats, extracting energy in the form of charged particles called electrons.

Proteins toward the end of the line — organized into five groups called complexes I, II, III, IV and V, and two mobile electron carriers, coenzyme Q10 and cytochrome c — harness the energy from those electrons to make ATP. Complexes I through IV shuttle the electrons down the line and are therefore called the electron transport chain, and complex V actually churns out ATP, so it is also called ATP synthase.

A deficiency in one or more of these complexes is the typical cause of a mitochondrial disease. (Mitochondrial diseases are sometimes named for a specific deficiency, such as complex I deficiency.) Coenzyme Q10 deficiency due to nuclear DNA mutations can present with proximal muscle weakness in isolation. In order to function correctly, the proteins of the oxidative phosphorylation pathway must be translated, imported into the mitochondria, and inserted into the inner mitochondrial membrane. Mutations in genes that affect these processes can also cause mitochondrial myopathy.

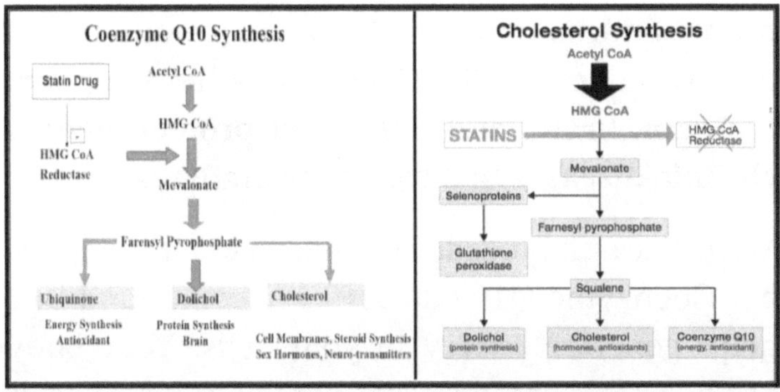

Picture Courtesy From - https://www.q10facts.com/coenzyme-q10-red-yeast-rice-heart-disease/

https://www.fxmedicine.com.au/blog-post/reaping-benefits-ubiquinol

Defects in genes related to the structure and dynamics of the mitochondria may also be involved in the development of myopathies. When a cell is filled with defective mitochondria, not only does it become deprived of ATP, but it can also accumulate a backlog of unused fuel molecules and oxygen, with potentially disastrous effects.

PQQ (Pyrroloquinoline Quinone) is a nootropic that supports memory, protects the brain and aids mitochondria by increasing their numbers.

It's found in a variety of foods, including dark chocolate and green tea, and are said to improve brain health, energy levels. SO, PQQ helps optimize the number of mitochondria. While, CoQ10 is a compound that works within the mitochondria, contributes to the production of energy in your cells. CoQ10 may help support the skin, brain, and lungs and protect against chronic diseases like cancer or diabetes.

More research is needed to understand its benefits, but Coenzyme Q10 (CoQ10) is a compound that helps generate energy in your cells. Your body produces less of it with age, but you can also get it from supplements or food.

Low levels of CoQ10 may be associated with diseases like cancer, diabetes, as well as neurodegenerative disorders. That said, the cause-effect relationship is unclear.

Here's what you need to know about its nine potential benefits and safety information.

CoQ10 is naturally found in the body, with the highest levels in the heart, liver, kidney, and pancreas. It helps generate energy in cells by making the antioxidant adenosine triphosphate (ATP), which is involved in cell energy transfer. It also serves as an antioxidant to protect cells against oxidative stress.

Ubiquinol is the reduced form of CoQ10, while ubiquinone is the oxidized form. The body is able to convert back and forth between these two forms.

Both variations exist in the body, but ubiquinol is the form that is found the most in blood circulation.

What does CoQ10 do for the body?

Oxidative stress can interfere with regular cell functioning and may contribute to many health conditions. Therefore, it is not surprising that some chronic diseases have also been associated with low levels of CoQ10.

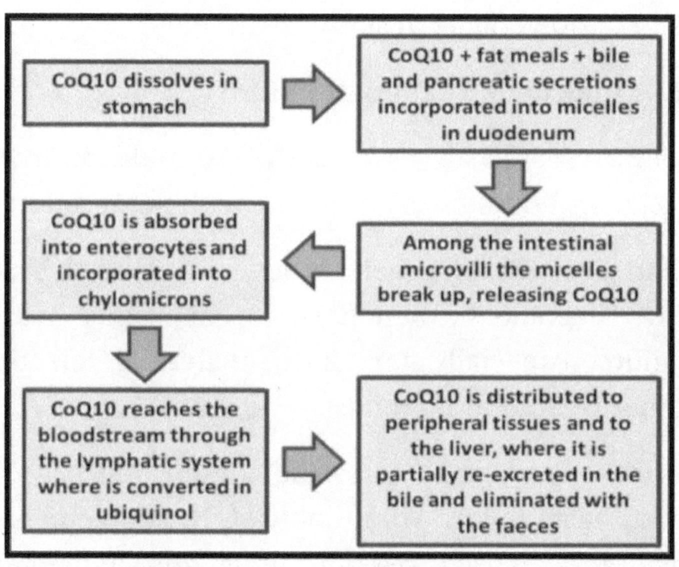

Picture Courtesy From - https://www.mdpi.com/2072-6643/13/5/1697

CoQ10 production decreases as you age. Thus, older people can be deficient in this compound.

Some other possible causes of low CoQ10 levels include:

- genetic defects in CoQ10 synthesis or utilization

- increased demands by tissues as a consequence of disease

- mitochondrial diseases

- oxidative stress due to aging

- side effects of statin treatments

1. It may help treat heart failure

Some research suggests that CoQ10 could improve treatment outcomes for people with heart failure.

One analysis of seven reviews concluded that CoQ10 could be beneficial for managing heart failure, especially for those unable to tolerate other treatment methods.

Another review of 14 studies found that people with heart failure who took CoQ10 supplements had a decreased risk of dying and a greater improvement in exercise capacity compared to those who took a placebo.

CoQ10 could also assist with restoring optimal levels of energy production, reducing oxidative damage, and improving heart function, all of which can aid the treatment of heart failure.

2. It could help with fertility

Female fertility decreases with age due to a decline in the number and quality of available eggs.

CoQ10 is directly involved in this process. As you age, CoQ10 production slows, making the body less effective at protecting the eggs from oxidative damage.

Supplementing with CoQ10 helps reverse this age-related decline in egg quality and quantity.

Similarly, male sperm is susceptible to oxidative damage, which may result in reduced sperm count, poor sperm quality, and infertility.

Several studies have concluded that supplementing with CoQ10 may improve sperm quality, activity, and concentration by increasing antioxidant protection.

3. It might help support healthy skin aging

Harmful elements like cellular damage or a hormonal imbalance can reduce skin moisture and protection from environmental aggressors, as well as thin the skin's layers.

According to some studies, applying CoQ10 directly to the skin may help reduce oxidative damage caused by UV rays, help decrease the depth of wrinkles, and promote antioxidant protection.

4. It could reduce headaches

Abnormal mitochondrial function can result in low energy in the brain cells and may contribute to migraine.

Since CoQ10 lives mainly in the mitochondria of the cells, it has been shown it may be beneficial for the treatment of migraine.

5. It could help with exercise performance

Abnormal mitochondrial function can reduce muscle energy, making it hard for muscles to contract efficiently and sustain exercise.

CoQ10 may help exercise performance by decreasing oxidative stress in the cells and improving mitochondrial function.

6. It may help with diabetes

Oxidative stress can induce cell damage. This can result in metabolic diseases like diabetes, as well as insulin resistance.

In a 2018 meta-analysis, CoQ10 has been suggested to improve insulin sensitivity and regulate blood sugar levels.

7. It might play a role in cancer prevention

According to some test-tube studies, CoQ10 could block the growth of cancer cells. Interestingly, people with cancer have been shown to have lower levels of CoQ10.

Some older studies suggest low levels of CoQ10 may be associated with a higher risk of certain types of cancer, including breast and prostate cancer. Newer studies have also suggested this with regard to lung cancer.

That said, the National Institutes of Health (NIH) Trusted Source states that CoQ10 is not of value as a cancer treatment, so more research needs to be conducted before a definitive claim can be made.

8. It may be good for the brain

Mitochondrial function tends to decrease with age, which can lead to the death of brain cells and contribute to conditions like Alzheimer's and Parkinson's.

Unfortunately, the brain is very susceptible to oxidative stress due to its high fatty acid content and oxygen demand.

This oxidative stress enhances the production of harmful compounds that could affect memory, cognition, and physical functions.

According to some animal studies in 2019Trusted Source and 2021Trusted Source, CoQ10 may reduce these harmful compounds, possibly slowing the progression of Alzheimer's and Parkinson's disease. However, more research on humans is needed.

9. It could protect the lungs

Increased oxidative damage in the lungs and poor antioxidant protection, including low levels of CoQ10, can result in lung diseases, such as chronic obstructive pulmonary disease (COPD) and asthma.

Dosage

CoQ10 comes in two different forms — ubiquinol and ubiquinone.

Current studies note that either ubiquinol or ubiquinone is acceptable as a supplement. No significant difference between the two was found in regard to absorption.

Because CoQ10 is a fat-soluble compound, its absorption is slow and limited. However, taking

CoQ10 supplements with food can help your body absorb it better than without food. Also, soft-gel capsules have been confirmed to absorb more efficiently than other forms of CoQ10.

Food sources of CoQ10

While you can easily consume CoQ10 as a supplement, it can also be found in some foods.

The following foods contain CoQ10:

Organ meats: heart, liver, and kidney

Some muscle meats: pork, beef, and chicken

Fatty fish: trout, herring, mackerel, and sardines

Legumes: soybeans, lentils, and peanuts

Nuts and seeds: sesame seeds and pistachios

Oils: soybean and canola oil

In addition to the foods listed above, some types of fruits, vegetables, dairy products, and cereals also contain CoQ10, though in much lower amounts.

Summary

CoQ10 is a fat-soluble, vitamin-like compound that seems to have some health benefits.

In particular, research suggests that it may help improve heart health and blood sugar regulation, protect against certain types of cancer, and reduce the frequency of migraine.

It may also reduce oxidative damage that leads to muscle fatigue, skin damage, and brain and lung diseases. However, more research is necessary to determine whether CoQ10 can help in these areas.

CoQ10 can be found as a supplement that seems well tolerated, but you should ask your doctor before trying it. You can also increase your intake through various food sources, including organ and muscle meats, oils, nuts, seeds, and legumes.

Ref From –

https://www.healthline.com/nutrition/coq10-dosage#takeaway

The Healthy Benefits of Coenzyme Q10 (CoQ10) with Macadamia Oil. https://www.drrobertyoung.com/post/the-benefits-of-coenzyme-q10-coq10-with-macadamia-oil

Mitochondrial Health May be the Key to Aging Well (and Having Kids) – Progressive Nutracare.

https://www.progressivenutracare.com/blogs/news/mitochondrial-health-may-be-the-key-to-aging-well-and-having-kids

https://www.mda.org/disease/mitochondrial-myopathies/causes-inheritance

Chapter – 19

Taurine and Anti-Aging

Taurine supplementation may be a viable solution to the problem of our cells manufacturing potentially hazardous by-products known as 'free radicals.' Some of these chemicals serve crucial biological activities, but excessive amounts can harm internal cell structures, reducing the cells' capacity to operate. The regulatory systems that contribute to maintaining a healthy balance of reactive oxygen species in the body deteriorate with age.

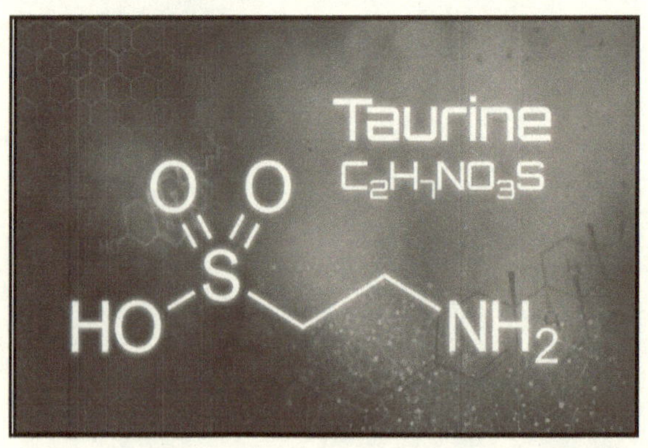

Picture Courtesy From - https://modernsportsnutrition.com/blogs/modern-sports-nutrition-blog/what-is-taurine-and-how-does-taurine-work

Aging is defined as a process that decreases the human body's capacity to withstand damage, disease, and stress. It affects physiological processes on many levels, including respiratory cycles, vision, blood pressure, and postural dynamics, and ultimately raises the risk of mortality and lowers fertility. Recent studies have revealed that the build-up of cellular and molecular damage causes aging.

This damage is mostly caused by reactions involving free radicals and other reactive oxygen species, as well as reactions involving sugars and reactive aldehydes and spontaneous mistakes in metabolic processes.

Oxygen metabolism and adenosine triphosphate (ATP) synthesis primarily produce reactive oxygen species (ROS) as a by-product. Excessive ROS are produced as a result of mitochondrial failure and antioxidant system degeneration in ageing, which causes cell dysfunction and death, for example cell cycle arrest, apoptosis and necrosis.

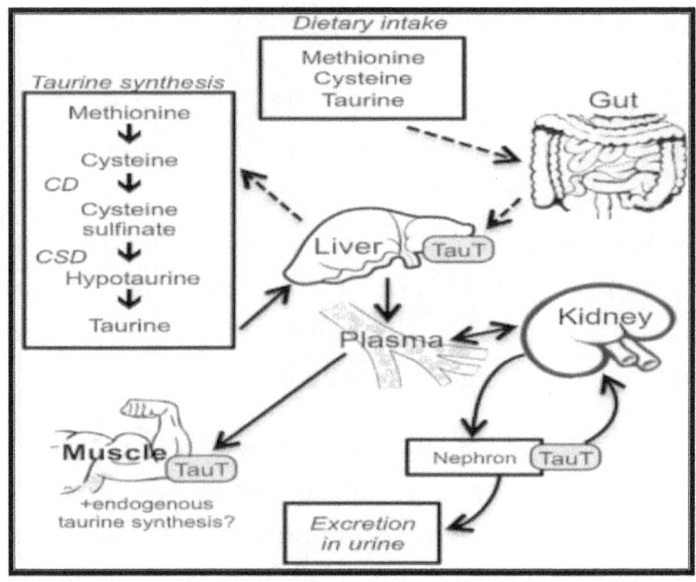

Picture Courtesy From - https://www.sciencedirect.com/science/article/abs/pii/S1357272515002010

Another major cellular process that contributes to whole body taurine content is the synthesis of taurine from cysteine. Cysteine is a semi-essential amino acid, which is present in the diet and can be synthesised from the **essential amino acid Methionine.**

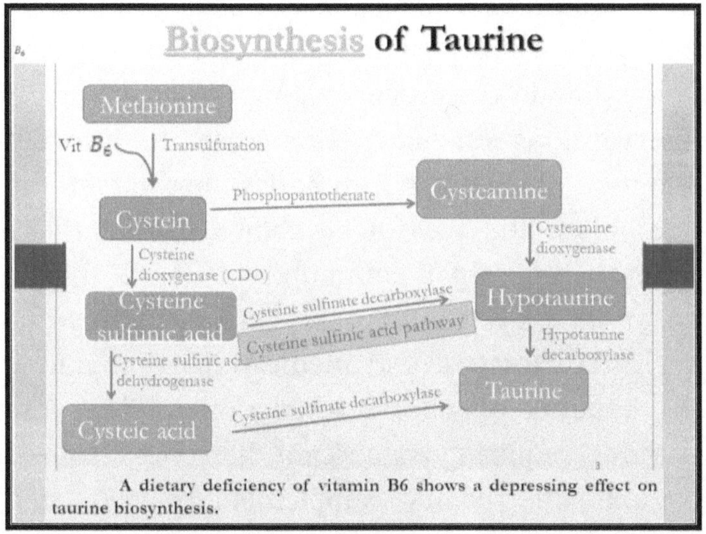

A dietary deficiency of vitamin B6 shows a depressing effect on taurine biosynthesis.

Picture Courtesy From - https://slideplayer.com/slide/1623242/

Along with metabolic alterations, abnormal oncogene activities, DNA damage accumulation, and telomere shortening, ROS also causes the multiplication of senescent cells as a stress response in aging.

These deficits play a role in the development of cancer, cardiovascular disease, metabolic disorders, exercise intolerance, and neurological disorders, which have led to an increased incidence of psychological disorders, including depression and anxiety, along with a gradually decreasing quality of life in the elderly population.

Dietary interventions such as caloric restriction, intermittent fasting, and time-restricted eating might slow down or reverse aging. Pharmacological interventions include telomerase to prevent telomere shortening, probiotics that target the microbiota and have been shown to effectively promote microbial community balance in the body in older adults and induce numerous health advantages, and medications that mimic the benefits of calorie restriction on long life without requiring significant lifestyle changes. Antioxidant dietary supplementation is also growing in popularity as an anti-aging therapy. However, their molecular pathways for oxidative stress defense and anti-aging actions are not fully understood.

Taurine, a naturally occurring sulfur-containing amino acid, has attracted significant attention in recent years due to its potential health benefits.

Evidence from both human and animal studies indicates that taurine may have beneficial cardiovascular effects, including blood pressure regulation, improved cardiac fitness, and enhanced vascular health. Its mechanisms of action and antioxidant properties make it also an intriguing candidate for potential anti-aging strategies.

Taurine (2-aminoethanesulfonic acid, also known as tauric acid) is a non-protein amino acid found in various animal tissues, especially in the brain, heart, and skeletal muscles. It is also present in several foods, such as meat, fish, dairy products, and energy drinks.

Taurine has been linked to antioxidant activity, anti-inflammatory effects, and blood pressure regulation, with significant implications for human health.

Early studies focused on its presence in animal tissues, where it is found in high concentrations in the brain, heart, and skeletal muscles. Later, in 1846, the English chemist Edmund Ronalds confirmed the presence of taurine in human bile. Taurine is detected in high concentrations in oxidative tissues, characterized by a high number of mitochondria, and in lower concentrations in glycolytic tissues.

Taurine has been extensively studied to determine its effects on human health. Regarding cellular function, taurine is primarily found in the intracellular fluid of many tissues, where it plays a vital role in several physiological processes. It acts as an osmolyte, regulating cell volume and maintaining cell integrity. In the liver, taurine is conjugated with bile acids, forming bile salts that aid in fat digestion and absorption in the intestines. These processes are crucial for lipid metabolism and absorption of fat-soluble vitamins.

Taurine has also been shown to be involved in calcium (Ca2+) signaling, modulation of ion channels, and neurotransmission, affecting neural excitability and synaptic transmission. Intriguingly, this amino acid exhibits essential antioxidant properties, protecting cells from oxidative and nitrosative stress by scavenging free radicals and reactive oxygen species (ROS).

Taurine is highly concentrated in the brain, and several studies indicate that it might act as a neurotransmitter or neuromodulator. It influences neurotransmitter release and receptor function, affecting cognitive processes, mood, behavior, memory, learning, and anxiety regulation.

Taurine has been thought to be essential for developing and surviving neural cells and protecting them under cell-damaging conditions; indeed, in the brain stem, taurine regulates vital functions, including cardiovascular control and arterial blood pressure. Its neuroprotective effects also involve reducing neuronal apoptosis and inflammation, making it a subject of interest in research on neurodegenerative diseases and brain injuries and offering benefits during stroke recovery. Premature infants are vulnerable to taurine deficiency because they lack some of the enzymes needed to synthesize cysteine and taurine.

Taurine plays a crucial role in cardiovascular physiology. Numerous studies have investigated the potential cardioprotective effects of taurine, focusing on its impact on blood pressure, cardiac contractility, and vascular function. It may help reduce blood pressure in individuals with hypertension and improve endothelial function, leading to enhanced vascular health. Its antioxidant properties may also reduce the risk of cardiovascular diseases such as atherosclerosis and heart failure.

The main cardiovascular effects of taurine are attributed to several underlying mechanisms. For instance, its modulation of ion channels, including Ca2+ and potassium (K+) channels, influences cardiac electrical activity and vascular tone. Its role in Ca2+ homeostasis also impacts myocardial contractility and relaxation. Additionally, the antioxidant properties of taurine, for which the exact underlying mechanisms remain unclear, might help protect against oxidative stress, a factor involved in the pathophysiology of cardiovascular disease.

Taurine has also been implicated in metabolic regulation, particularly in relation to glucose and lipid metabolism. Various studies indicate that taurine might help improve insulin sensitivity, making it beneficial for individuals with type 2 diabetes (T2D) or those at risk of developing the condition. A recent preclinical study has shown that taurine can rescue pancreatic β-cell stress by stimulating α-cell transdifferentiation.

Additionally, taurine may aid in reducing triglyceride levels and improving lipid profiles, potentially lowering the risk of cardiovascular diseases and metabolic syndrome.

The endothelium, a single layer of cells lining the blood vessels, plays a crucial role in vascular health. Taurine has been shown to improve endothelial function by promoting nitric oxide (NO) production and reducing endothelial dysfunction. Enhanced endothelial function contributes to better vascular relaxation, reduced inflammation, and improved blood flow, which may benefit cardiovascular health and reduce the risk of atherosclerosis and cardiovascular events.

Levels of taurine have been shown to decline as we age, and offsetting this loss with a taurine supplement might delay the development of age-related health problems. Indeed, as shown in a *recently published science paper*, when mice received taurine supplements, their lifespans increased by approximately 10% compared to the control group. Mice in the taurine group also seemed healthier, with improvements in muscle endurance and strength. Researchers fed mice between 15 and 30 mg of taurine per day, depending on their age.

Taurine was also shown to shape mice's gut microbiota and positively affect intestinal homeostasis restoration, suggesting that it could be harnessed to re-establish a normal microenvironment and treat or prevent gut dysbiosis.

Sirtuins are a family of proteins that possess either mono-ADP-ribosyltransferase or deacetylase activity. They regulate many signalling pathways, mostly connecting them with the organism's metabolic state.

Their expression is decreased with age and their activation or overexpression is associated with increased longevity. Taurine was shown to activate cytoplasmic SIRT1 in the liver, heart, and brain.

Taurine may strengthen your body's antioxidant defenses and reduce your risk of diabetes, high blood pressure and cardiovascular disease. It is naturally created in a few body tissues, particularly the liver, and plays an essential role in the integrity of the central nervous system, immunity, vision and fertility.

The inclusion of taurine in the diet may provide a widely available, inexpensive, low-risk method of preventing aging with results that are better compared to currently available, expensive anti-aging therapies. Red meat, organ meats, chicken, turkey, eggs, shellfish like scallops, mussels and clams, and modern products like fizzy drinks all contain significant taurine concentrations. **Taurine supplements cannot work wonders independently; they can only help when additional efforts are made, such as leading a healthy lifestyle with a balanced diet and regular exercise.**

Ref From –

https://www.ncbi.nlm.nih.gov/pmc/articles/
PMC10328712/

https://www.ncbi.nlm.nih.gov/pmc/articles/
PMC10574552/

Santulli, G., Santulli, G., Kansakar, U., Varzideh, F., Mone, P., Jankauskas, S., & Lombardi, A. (2023). Functional Role of Taurine in Aging and Cardiovascular Health: An Updated Overview. Nutrients, 15(19), 4236.

Chapter – 20

NIACIN (B3) and Anti-Aging

Did you know there are eight kinds of Vitamin B and one of them can slow down the aging process?

These B-complex vitamins are made up of thiamine (B1), riboflavin (B2), niacin (B3), pantothenic acid (B5), pyridoxine (B6), biotin (B7), folate (B9), and cobalamin (B12).

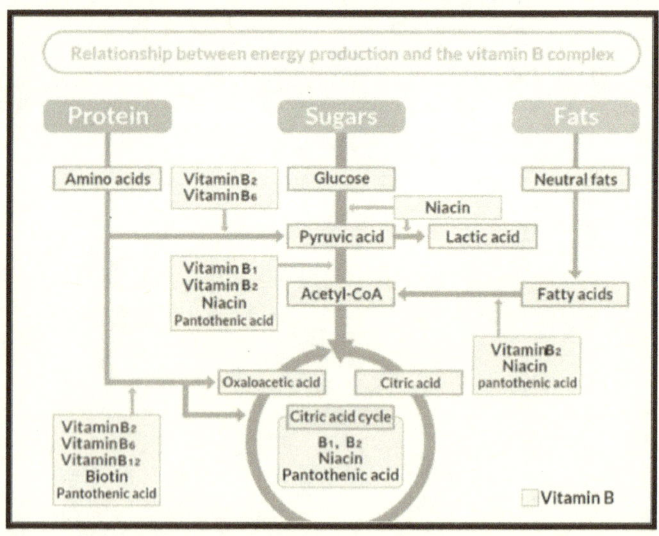

Picture Courtesy From - https://www.otsuka.co.jp/en/nutraceutical/about/nutrition/sports-nutrition/essential-nutrients/vitaminbcomplex.html

Vitamin B3 —or niacin— comprises three chemical forms called nicotinic acid, nicotinamide, and nicotinamide riboside, which play a big part in your metabolism and aging.

Nicotinic acid, nicotinamide, and nicotinamide riboside are the building blocks of NAD (nicotinamide adenine dinucleotide), which is critical in optimal cell regeneration. Cell regeneration keeps your body young.

The body's creation of NAD depletes as we age, so increasing the intake of niacin as we age is important to maintain healthy energy metabolism and, in turn, cell regeneration.

Niacin is a precursor to NAD, which is vital in cell regeneration. Since NAD creation depletes as you age, you need to increase your intake of niacin to maintain healthy cell regeneration.

Although the study showed niacin being absorbed into the bloodstream, your body cannot store vitamin B3 for long periods of time, and excess vitamin B3 not absorbed by the body passes through as waste, so you need to restore these vitamins through food or supplements constantly.

Niacin is a form of water-soluble vitamin B3 and has been known since American biochemist Conrad Elvehjem identified it in 1937. Niacin was originally called nicotinic acid because it can be created by the oxidation of nicotine with nitric acid.

However, people know nicotine to be the addictive chemical in tobacco, so the name niacin was adopted, which comes from the words **NIcotinic ACid vitamIN**.

There has been considerable interest in supplements that have the ability to increase NAD+ levels in recent years, and niacin is one such supplement that can do this. NAD+ biology plays an important role in energy metabolism.

Aging is a process that seems impossible to stop, but there might be ways you can slow it down. Vitamin B3, also known as niacin, is a vitamin that may have anti-aging benefits and could be important to add to your diet if you want to stay healthy in your later years.

Vitamin B3 is a water-soluble vitamin that's part of the vitamin B complex. It's found in both plant and animal foods and is important for the body to function properly.

Some of the benefits of vitamin B3 include:

- helping to maintain healthy skin, hair, and nails

- regulating cholesterol levels

- supporting energy production

- helping to process food and nutrients

So, what does this have to do with anti-aging?

Well, vitamin B3 is also thought to slow down the aging process. Some scientific evidence shows that vitamin B3 may help slow the aging process. One study showed that when rats were given vitamin B3 supplements, they had less age-related damage in their brains and eyes. Another study found that vitamin B3 improved the appearance of skin wrinkles in people who took it for 12 weeks.

While these studies are promising, more research is needed to say for sure whether or not vitamin B3 can help slow down aging,

If you're interested in adding vitamin B3 to your diet, there are plenty of ways to do it. You can find vitamin B3 in animal and plant foods, so it's easy to include in your diet. Some good sources of vitamin B3 include:

- fish such as salmon, trout, and tuna

- beef

- poultry

- eggs

- legumes

- nuts

- seeds

- whole-grain bread and cereals

- green leafy vegetables

While there are overall benefits to taking vitamin B3, a few key benefits of taking vitamin B3 include anti-aging. These include:

One of the main benefits of vitamin B3 is that it can help reduce wrinkles and other signs of aging.

This is because vitamin B3 helps increase collagen production, which helps keep skin looking smooth and elastic.

One vitamin B3 anti-aging property you might have overlooked is its antioxidant function and ability to combat free radicals.

Free radicals can damage many cells in your body, including skin cells. Vitamin B3 can, therefore, help reduce age spots, which are often a by-product of free radical damage.

Another benefit of vitamin B3 for anti-aging is that it helps to improve skin health. Vitamin B3 can help reduce dryness and flakiness and enhance the appearance of redness and blemishes. It can also help protect the skin because vitamin B3 helps reduce inflammation and improve blood circulation, thereby giving skin a healthy glow.

Vitamin B3 has also been shown to help reduce the risk of age-related diseases, such as heart disease and dementia. It provides this benefit by helping to keep cells healthy and functioning properly, in part due to its function as an antioxidant. These diseases often become more common as we age, so adding vitamin B3 to your diet could help reduce your risk of developing them.

As we age, it's common to start feeling tired and sluggish. This might be due to several things, including the natural aging process and health conditions. Vitamin B3 can help improve energy levels by helping the body produce energy from food more effectively.

Furthermore, vitamin B3 has also been shown to help improve mood. This is likely due to its ability to reduce inflammation and oxidative stress. Both of these can lead to feelings of anxiety and depression.

As with any supplement, it is important to take the correct amount of vitamin B3, as consuming too much vitamin B3 can cause certain side effects.

These can include nausea, vomiting, and diarrhea. If you are considering taking vitamin B3 supplements, it is important to speak with your doctor to ensure that they are the right supplement for you and that you are not at risk of any adverse effects.

Ref From –

https://www.endur.com/blogs/blog/how-niacin-can-slow-down-the-aging-process

https://pubmed.ncbi.nlm.nih.gov/32386566/

https://www.healthycell.com/blogs/articles/do-vitamin-b3-anti-aging-supplements-work#:~:text=Aging%20is%20a%20process%20that,healthy%20in%20your%20later%20years.

https://www.lifespan.io/news/niacin-increases-nad-significantly-in-human-trial/

Chapter – 21

Methyl cobalamin (B12) and Anti-Aging

The B12 vitamin is an essential part of good health. It's naturally found in foods you eat, including meats, milk, eggs, and other dairy products and plays a vital role in the production of red blood cells. B12 also gives your brain the energy it needs to function properly.

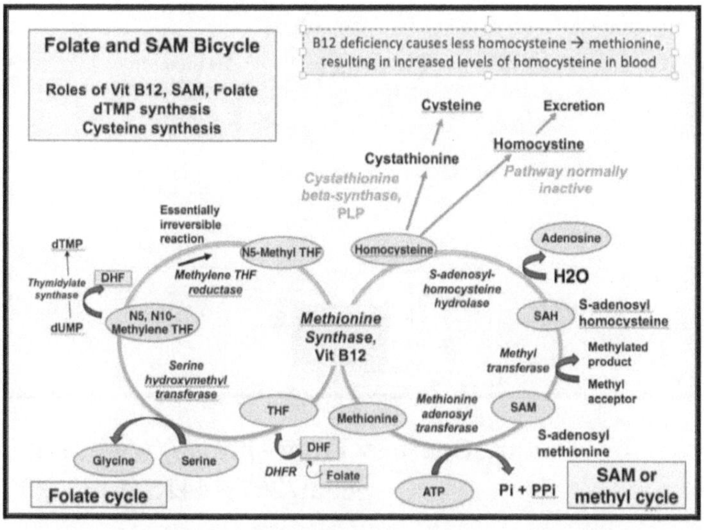

Picture Courtesy From - https://uw.pressbooks.pub/fmrbiochemistry/chapter/main-body-13/

It's found naturally in animal products, but also added to certain foods and available as an oral supplement or injection.

Vitamin B12 has many roles in your body. It supports the function of your nerve cells and is needed for red blood cell formation and DNA synthesis. Vitamin B12 is vital in helping your body produce red blood cells.

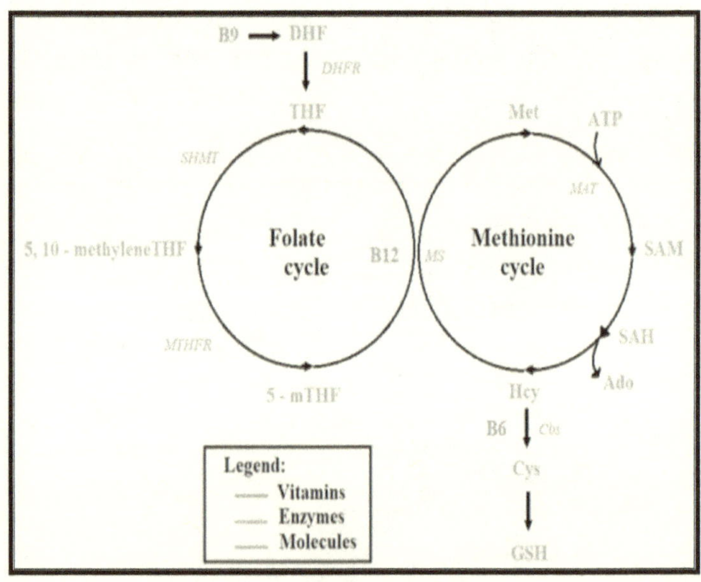

Picture Courtesy From - https://www.mdpi.com/2073-4425/14/3/709

Low vitamin B12 levels reduce red blood cell formation and prevent them from developing properly. Healthy red blood cells are small and round, whereas they become larger and typically oval in cases of vitamin B12 deficiency.

Due to this larger and irregular shape, the red blood cells are unable to move from the bone marrow into the bloodstream at an appropriate rate, causing megaloblastic anemia. When you have anemia, your body doesn't have enough red blood cells to transport oxygen to your vital organs. This can cause symptoms like fatigue and weakness.

Vitamin B12 (cobalamin) is an essential nutrient for humans and animals. Metabolically active forms of B12-methylcobalamin and 5-deoxyadenosylcobalamin are cofactors for the enzyme's methionine synthase and mitochondrial methyl malonyl-CoA mutase. Malfunction of these enzymes due to a scarcity of vitamin B12 leads to disturbance of one-carbon metabolism and impaired mitochondrial function. A significant fraction of the population (up to 20%) is deficient in vitamin B12, with a higher rate of deficiency among elderly people.

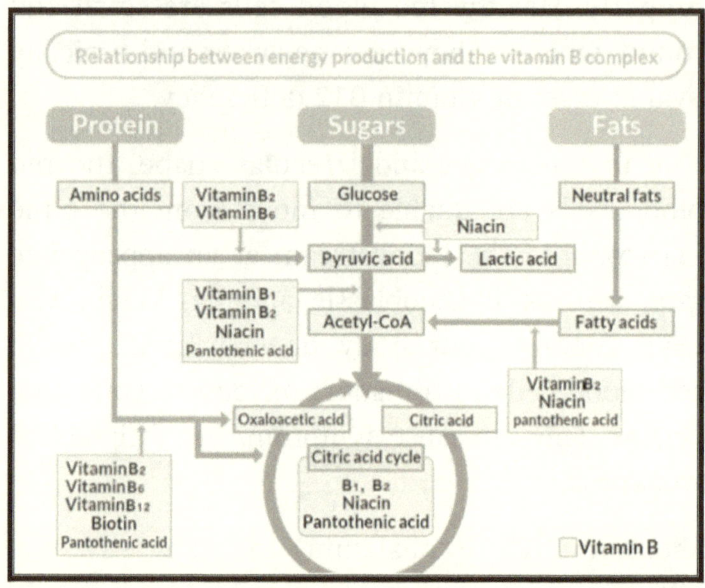

Picture Courtesy From - https://www.otsuka.co.jp/en/nutraceutical/about/nutrition/sports-nutrition/essential-nutrients/vitaminbcomplex.html

B12 deficiency is associated with numerous hallmarks of aging at the cellular and organismal levels. Cellular senescence is characterized by high levels of DNA damage by metabolic abnormalities, increased mitochondrial dysfunction, and disturbance of epigenetic regulation.

Vitamin B12 is the generic name for a group of cobalamin that play an important role in physiological processes. Recent studies elucidate that vitamin B12 may be involved in several pathological processes: change in DNA methylation, aberrant protein post-translational modifications, change in microbial composition in the gut, change in tumorigenesis probability, and modulation of inflammation.

We discuss how these B12-associated processes may contribute to the acceleration of aging and senescence and to inflammation. This is in addition to the increased risk of age-related diseases, such as cognitive decline, cardiovascular diseases, osteoporosis, and oxidative stress, which are alleviated by supplementation of vitamin B12.

Defects in DNA methylation in cells lead to impaired gene repression and aberrant expression, which disrupt their homeostasis, increase tumor occurrence, and promote cellular aging.

Methylcobalamin vs. Cyanocobalamin: What's the Difference?

Both cyanocobalamin and methylcobalamin are forms of vitamin B12. Cyanocobalamin is synthetic while methyl cobalamin is natural. However, the body can convert synthetic forms into natural ones.

Vitamin B12, also known as cobalamin, is an important water-soluble vitamin in red blood cell production, brain health, and DNA synthesis. A deficiency in this key vitamin can cause serious symptoms, including fatigue, nerve damage, digestive issues, and neurological problems like depression and memory loss.

Synthetic vs. Natural

Vitamin B12 supplements are typically derived from two sources: cyanocobalamin or methylcobalamin. Both are nearly identical and contain a cobalt ion surrounded by a corrin ring.

However, each has a different molecule attached to the cobalt ion. While methylcobalamin contains a methyl group, cyanocobalamin contains a cyanide molecule.

Cyanocobalamin is a synthetic form of vitamin B12 that's not found in nature. It's used more frequently in supplements, as it's considered more stable and cost-effective than other forms of vitamin B12. When cyanocobalamin enters your body, it's converted into either methylcobalamin or adenosylcobalamin, which are the two active forms of vitamin B12 in humans. Unlike cyanocobalamin, methylcobalamin is a naturally occurring form of vitamin B12 that can be obtained through supplements, as well as food sources like fish, meat, eggs, and milk.

Both methylcobalamin and cyanocobalamin can be converted to other forms of vitamin B12.

When you ingest cyanocobalamin, it can be converted to both of the active forms of vitamin B12, methylcobalamin and adenosylcobalamin.

Much like methylcobalamin, adenosylcobalamin is essential to many aspects of your health.

It's involved in the metabolism of fats and amino acids, as well as the formation of myelin, which creates a protective sheath around your nerve cells. Deficiencies in both forms of vitamin B12 can increase your risk of neurological issues and adverse side effects.

If you think you may have a vitamin B12 deficiency, talk to your doctor to determine the best course of treatment.

Ref From –

https://www.ncbi.nlm.nih.gov/pmc/articles/PMC11084641/

https://www.healthline.com/nutrition/vitamin-b12-benefits#more-energy

https://grossmanmed.com/b12-injections-for-anti-aging/

https://www.healthline.com/nutrition/
methylcobalamin-vs-cyanocobalamin#bottom-
line

Simonenko, S. Y., Bogdanova, D. A., &
Kuldyushev, N. A. (2024). Emerging Roles
of Vitamin B12 in Aging and Inflammation.
International Journal of Molecular Sciences.
https://doi.org/10.3390/ijms25095044

What are the benefits of B Vitamins? | ActiKid,
active, avocado and more | ActiKid Blog blog.
https://actikid.com/blogs/news/what-are-the-
benefits-of-b-vitamins

Methylcobalamin And Cyanocobalamin -
What's The Difference? - Purality Health®
Liposomal Products. https://puralityhealth.com/
methylcobalamin-and-cyanocobalamin-whats-
the-difference/

Chapter – 22

Glynac (Glycine and N-Acetylcysteine) and Anti-Aging

Supplementing **Glycine and N-Acetylcysteine (GlyNAC)** in Older Adults Improves Glutathione Deficiency, Oxidative Stress, Mitochondrial Dysfunction, Inflammation, Physical Function, and Aging Hallmarks: A Randomized Clinical Trial.

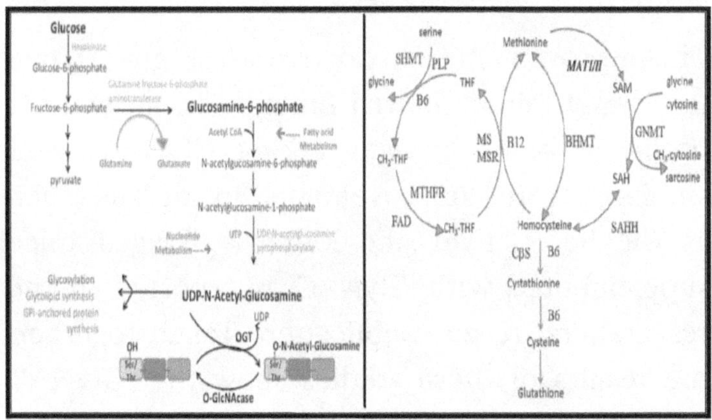

Picture Courtesy From -https://www.researchgate.net/figure/The-synthesis-of-Uridine-diphospho--N-acetylglucosamine-UDP-GlcNAc-from-glucose_fig1_299857510

https://www.researchgate.net/figure/Schematic-illustration-of-one-carbon-metabolism-pathways-where-vitamins-B6-B12-and-folic_fig1_265128704

Elevated oxidative stress (OxS), mitochondrial dysfunction, and hallmarks of aging are identified as key contributors to aging, but improving/reversing these defects in older adults is challenging. In some studies, deficiency of the intracellular antioxidant glutathione (GSH) could play a role, and it reported that supplementing GlyNAC (combination of glycine and N-acetylcysteine [NAC]) in aged mice improved GSH deficiency, OxS, mitochondrial fatty-acid oxidation (MFO), and insulin resistance (IR).

The improvements in oxidative stress, glutathione levels and mitochondrial function in the muscle tissue of older humans taking GlyNAC were similar to the improvements in organs such as the heart, liver and kidneys of aged mice supplemented with GlyNAC as reported in the researchers' recent publication. Taken together, the results of these studies show that GlyNAC supplementation can improve these defects in many different organs of the body.

A randomized, double-blind human clinical trial conducted by researchers at Baylor College of Medicine reveals that supplementation with GlyNAC—a combination of glycine and N-acetylcysteine—improves many age-associated defects in older humans and powerfully promotes healthy aging.

Published in the Journal of Gerontology: Medical Sciences, the study shows that older humans taking GlyNAC for 16 weeks improved many characteristic defects of aging. These include oxidative stress, glutathione deficiency, and multiple aging hallmarks affecting mitochondrial dysfunction, mitophagy, inflammation, insulin resistance, endothelial dysfunction, genomic damage, stem cell fatigue, and cellular senescence. These were associated with muscle strength, gait speed, exercise capacity, waist circumference, and blood pressure improvements.

Many people aspire to live longer, healthier lives. A researcher at Baylor College of Medicine, Dr. Rajagopal Sekhar, associate professor of medicine in the section of endocrinology, diabetes, and metabolism, has been studying aging for the past 20 years to understand its biological underpinnings better and develop nutritional strategies to promote healthy aging.

In today's issue of the journal Nutrients, Sekhar's team reports the results of a study in mice. This provides proof of concept that supplementing GlyNAC—a combination of glycine and N-acetylcysteine as precursors of the natural antioxidant glutathione—can increase lifespan and improve multiple key age-associated defects.

Sekhar has studied natural aging in older humans and mice for over two decades. His work contributes to a better understanding of how glutathione deficiency, increased oxidative stress, mitochondrial dysfunction, and multiple additional hallmark defects of aging contribute to the aging process and how they can be reversed with GlyNAC supplementation.

The researchers housed the mice under stable environmental conditions. When mice reach age 65-weeks, they typically begin to show a drop in glutathione levels and develop mitochondrial dysfunction and oxidative stress. At this age, Sekhar's team switched the diets of half of the mice to receive GlyNAC, and the other half continued their diet without the supplement. Except for the GlyNAC supplementation, the diets of both groups were the same in terms of protein, fat, and carbohydrates. Then, the researchers let the mice continue aging undisturbed and recorded how long they lived.

"We were excited to find that the mice that received GlyNAC lived 24% longer than those that did not receive GlyNAC," Sekhar said. "We next wanted to understand how GlyNAC works."

For this, the team conducted a second study in aged mice to investigate mitochondrial dysfunction, glutathione levels, oxidative stress, and other hallmarks of aging, specifically in the animals' heart, liver, and kidneys. The researchers determined the effect of supplementing GlyNAC on these cellular defects. The researchers chose these organs because they perform vital functions in the body.

In aged mice, the team found that all three organs had glutathione deficiency, oxidative stress, mitochondrial dysfunction, abnormal mitophagy (difficulty in disposing of damaged mitochondria), impaired nutrient sensing, and genomic damage. GlyNAC supplementation improved and corrected all these defects in the old mice. The researchers propose that it is the improvement of these fundamental biological defects that contribute to the animals' longer life.

"But mice are not humans. Could this happen to people? There is published evidence from our human studies showing that GlyNAC supplementation improves similar defects in people," Sekhar said.

For instance, in a pilot human clinical trial in older adults, Sekhar's group showed that taking GlyNAC for 24 weeks improved many characteristic defects of aging, including glutathione deficiency, oxidative stress, mitochondrial dysfunction, inflammation, insulin resistance, endothelial dysfunction, body fat, and genomic damage.

Sekhar thinks that the glycine and NAC (from GlyNAC), combined with glutathione (that GlyNAC helps generate), provide the benefits described above and he calls this combination the 'power of 3.'

Ref From –

https://pubmed.ncbi.nlm.nih.gov/35975308/

https://www.bcm.edu/news/glynac-supplementation-reverses-aging-hallmarks-in-aging-humans

https://www.news-medical.net/news/20220307/GlyNAC-supplementation-can-increase-lifespan-and-improve-multiple-age-associated-defects.aspx

Fernández-Lázaro, D., Fernández-Lázaro, D., Domínguez-Ortega, C., Busto, N., Santamaría-Peláez, M., Roche, E., Gutiérez-Abejón, E., Mielgo-Ayuso, J., & Mielgo-Ayuso, J. (2023). Influence of N-Acetylcysteine Supplementation on Physical Performance and Laboratory Biomarkers in Adult Males: A Systematic Review of Controlled Trials. Nutrients, 15(11), 2463.

GlyNAC supplementation reverses aging hallmarks in aging humans. https://medicalxpress.com/news/2022-08-glynac-supplementation-reverses-aging-hallmarks.html

GlyNAC supplementation extends life span in mice – CLINICALNEWS.ORG. https://clinicalnews.org/2022/03/07/glynac-supplementation-extends-life-span-in-mice/amp/

Chapter – 23

L-Arginine and Anti-Aging

L-arginine is one of the most metabolically versatile amino acids. In addition to its role in the synthesis of nitric oxide, l-arginine serves as a precursor for synthesizing polyamines, proline, glutamate, creatine, agmatine, and urea. Several human and experimental animal studies have indicated that exogenous l-arginine intake has multiple beneficial pharmacological effects when consumed as dietary.

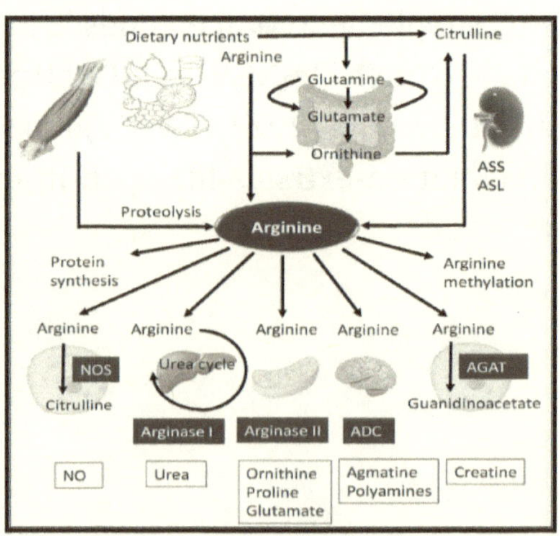

Picture Courtesy From - https://www.mdpi.com/2072-6643/11/7/1452

Such effects include:

- a reduction in the risk of vascular and heart diseases

- a reduction in erectile dysfunction

- an improvement in immune response

- inhibition of gastric hyperacidity

L-arginine is a semi-essential amino acid that is not only involved in protein synthesis but is also a single substrate for endothelial nitric oxide synthase (eNOS) to produce the essential vasoprotective molecule nitric oxide (NO). Depletion of L-arginine causes eNOS dysfunction in cultured endothelial cells. Hence, L-arginine supplementation has been widely used in many physiological and pathological conditions to improve health status or treat cardiovascular diseases.

L-arginine is a basic natural amino acid. Hedin discovered its presence in mammalian proteins in 1895. **L-arginine is engaged in several metabolic pathways within the human body. It is an essential component of the urea cycle, the only pathway in mammals that allows the elimination of toxic ammonia from the body.**

L-arginine is the only substrate in the biosynthesis of NO, which plays critical roles in diverse physiological processes in the human body, including neurotransmission, vasorelaxation, cytotoxicity, and immunity.

The primary tissue in which endogenous l-arginine synthesis occurs is the kidney, where l-arginine is formed from citrulline, which is released mainly by the small intestine. The liver is also capable of synthesizing considerable amounts of l-arginine; however, this is completely reutilized in the urea cycle so that the liver contributes little or not at all to plasma arginine flux.

L-Arginine and the cardiovascular system

In the cardiovascular system, exogenous l-arginine causes a rapid reduction in systolic and diastolic pressures when infused into healthy humans and patients with various forms of hypertension. Furthermore, oral l-arginine supplementation attenuates platelet reactivity and improves endothelial function in animal models of hypercholesterolemia and atherosclerosis. Clinical studies of l-arginine in humans have also been highly positive in improving endothelial dysfunction and preventing restenosis after balloon angioplasty. An excellent review of the clinical pharmacology of l-arginine, particularly in the cardiovascular system, has been provided by Boger and Bode Boger.

L-Arginine and sexual function

Arginine is required for normal spermatogenesis. Over 50 years ago, researchers found that feeding an arginine-deficient diet to adult men for 9 days decreased sperm counts by 90% and increased the percentage of non-motile sperm approximately 10-fold.

Other effects of L-arginine

Other than the benefits in the above stated conditions, l-arginine has been demonstrated to improve peripheral circulation, renal function, and immune function. It also possesses anti-stress and adaptogenic capabilities. l-Arginine stimulates the release of growth hormone as well as the release of pancreatic insulin and glucagon and pituitary prolactin.

Summary as per few Researchers:

L-arginine is a semi-essential amino acid involved in protein synthesis and is the substrate for nitric oxide synthase (NOS) to produce the vascular protective nitric oxide (NO) released from the endothelial cells (Wu et al., 2009). Since decreased bioavailability of NO or deficiency of NO production promotes development of cardiovascular diseases and chronic kidney diseases, and is highly associated with aging (Schmitt and Melk, 2017; Donato et al., 2018), supplementation of L-arginine has been proposed to increase endothelial NO bioavailability and to improve health status in the young as well as in elderly population or as an adjunct therapeutic modality to treat patients with cardiovascular diseases (Creager et al., 1992). However, mixed results on the therapeutic effects of L-arginine supplementation in either experimental models or clinical studies are reported and there is continuous debate on whether L-arginine shall be supplemented or shall be avoided and in which context of diseases (Dioguardi, 2011; Nogiec and Kasif, 2013; Hadi et al., 2019; Nitz et al., 2019). While some studies demonstrate improvement of cardiovascular functions or reduced risks (Rector et al., 1996; Maxwell et al., 2000; Dong et al., 2011), numerous other studies with long term L-arginine supplementation showed either no sustained effects (Sato et al., 2000; Meirelles and Matsuura, 2018; Rodrigues-Krause et al., 2018) or harmful effects and even increased mortality in patients with cardiovascular disease

(Chen et al., 2003; Schulman et al., 2006; Wilson et al., 2007).

Ref From –

https://www.sciencedirect.com/science/article/pii/S2090123210000573

https://www.ncbi.nlm.nih.gov/pmc/articles/PMC4069264/

https://www.ncbi.nlm.nih.gov/pmc/articles/PMC7851093/

Gad, M. Z. (2010). Anti-aging effects of l-arginine. Journal of Advanced Research. https://doi.org/10.1016/j.jare.2010.05.001

L-Arginine: the best anti-aging remedy – IronMag Bodybuilding & Fitness Blog. https://www.ironmagazine.com/2015/l-arginine-the-best-anti-aging-remedy/

Adebayo, A., Varzideh, F., Wilson, S., Gambardella, J., Eacobacci, M., Jankauskas, S., Donkor, K., Kansakar, U., Trimarco, V., Mone, P., Lombardi, A., Santulli, G., & Santulli, G. (2021). L-Arginine and COVID-19: An Update. Nutrients, 13(11), 3951.

Chapter – 24

Melatonin and Anti-Aging

Well-known for its role in the sleep-wake cycle, melatonin is now being studied for its pro-aging properties.

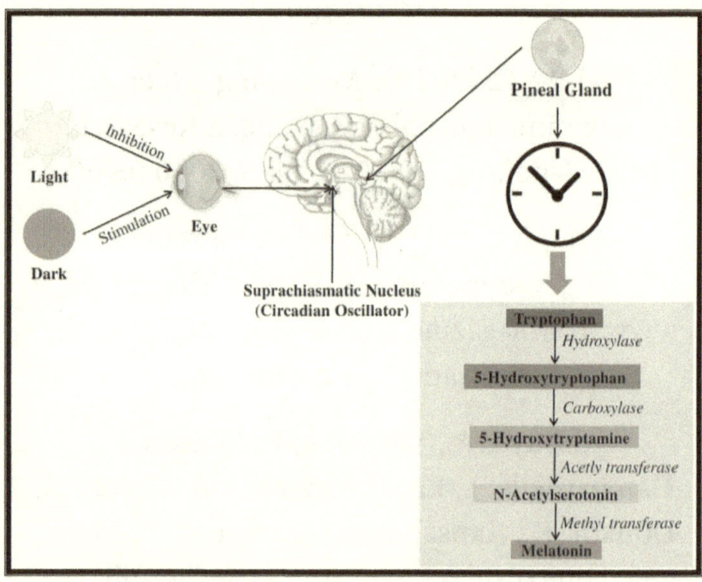

Picture Courtesy from - https://www.researchgate.net/figure/Regulation-of-melatonin-synthesis-by-the-pineal-gland-to-control-circadian-rhythms-in-the_fig1_333311572

While melatonin is best known for regulating the sleep-wake cycle, current research has entered a new realm: investigating the hormone's potential pro-aging properties. Melatonin is an endogenous indoleamine that is synthesized not only in the pineal gland (endocrine gland located in the posterior area of the cranial fossa in the brain in charge of producing melatonin and serotonin-derived hormones crucial in the modulation of sleep patterns in both circadian and seasonal cycles) but also perhaps in each individual cell, specifically at the mitochondrial level (Yanar et al., 2019).

Studies suggest that melatonin possesses potent antioxidant properties. It acts as a scavenger of highly reactive molecules known as free radicals, which are notorious for causing oxidative damage, which is implicated in the aging process.

Melatonin's antioxidant effects may help alleviate oxidative stress and safeguard cells from harm, essentially working against aging.

In addition, melatonin regulates inflammation, mitochondrial function, and the immune system, all of which play a role in aging. Melatonin may also aid in producing other hormones, including growth hormone, which is essential for tissue repair and regeneration.

Research suggests that melatonin production tends to decline with age. This reduction in melatonin levels may result in the body's reduced ability to counteract oxidative stress and protect against free radical damage.

As a result, the skin cells may become more susceptible to damage from environmental factors, such as UV radiation from the sun, pollution, and other sources of oxidative stress. This leads to accelerated skin aging, including the formation of wrinkles.

Though melatonin has strong antioxidant properties and may counteract the aging process, its ability to directly slow the aging process in humans has not been definitively proven.

Still, some animal studies have shown that melatonin supplementation can extend lifespan and improve health span in some animal models. For instance, a 2002 study found that melatonin can extend the life of fruit flies by 33%. But these findings may not directly translate to humans.

Melatonin is a powerful antioxidant. But it's important to note that there's likely no single "best" antioxidant for pro-aging.

Antioxidants are a diverse group of compounds that work together to neutralize harmful free radicals and reduce oxidative stress, associated with aging and various age-related diseases.

Other antioxidants, such as vitamin C, vitamin E, and beta-carotene, play vital roles in the body's defense against oxidative stress.

As humanity continues to search for the elusive fountain of youth, one intriguing area of research revolves around the potential pro-aging properties of melatonin.

Scientists are studying how melatonin impacts the inner workings of cells and the body's oxidative stress levels, which may help uncover its potential effect on graceful aging.

What are the pro-aging benefits of melatonin?

Some of the reported pro-aging benefits of melatonin include:

Antioxidant properties

Melatonin acts as a potent antioxidant, which can help neutralize harmful free radicals and reduce oxidative stress. Oxidative stress is a process that damages cells and has been implicated in the aging process. By reducing oxidative stress, melatonin may help protect against cellular damage and counteract the aging process.

Sleep regulation

Melatonin is well-known for regulating the sleep-wake cycle, and adequate sleep is essential for overall health and well-being. Good quality sleep is linked to improved cognitive and immune function and overall health, which can have potential pro-aging effects.

Mitochondrial health

Mitochondria are the energy-generating powerhouses of cells, and their dysfunction plays a role in aging. Melatonin has been shown to protect against mitochondrial damage and improve mitochondrial function.

Potential impact on collagen

As we age, our skin loses elasticity and collagen. Collagen is a protein that provides structural support to the skin, bones, tendons, and other connective tissues. It plays a critical role in maintaining the integrity and elasticity of tissues.

Melatonin is known for helping some people achieve a better night's sleep, which is why it is nicknamed "the sleep hormone."

While melatonin regulates our internal body clock, that's just the beginning of its health-promoting benefits. Based on extensive research, scientists have discovered that this hormone has beneficial effects on everything from heart disease and diabetes to bone health and obesity. Best of all, emerging science now suggests that it may protect our genetic material and guard against age-related disease and decline.

Preclinical studies found that melatonin increased the life span of animals by up to **20%**—prolonging their youthful character in the process. Scientific evidence suggests that melatonin is crucial in various metabolic functions, including antioxidant and neuroprotection, anti-inflammatory defense, and immune system support.

Because melatonin production reduces with age, experts believe its decline contributes to both the aging process and a generalized deterioration of health.

A growing body of evidence reveals how melatonin plays such a major role in combating the aging process.

Beyond Sleep: Ways Melatonin Attacks Aging Factors

Melatonin is involved in regulating our internal body clock. However, scientists are discovering that this hormone has beneficial effects on everything from heart disease and diabetes to bone health and obesity. Emerging science now suggests that it may protect our genetic material and guard against age-related decline.

Scientifically reviewed by: Dr. Amanda Martin, DC, in August 2023. Written by: Claudia Kelley, PHD, RD, CDE.

Antioxidant Defence—Combat Free Radical Damage While You Sleep

Since its discovery over 50 years ago, melatonin has demonstrated itself as a functionally diverse molecule, with its antioxidant properties being amongst its most well-studied attributes. Since then, a vast amount of experimental research has revealed its vital role in the body's defense against numerous cell-damaging free radicals—and for good reason.

Melatonin has been found to possess **200%** more antioxidant power than vitamin E. It is also superior to glutathione and vitamins C and E in reducing oxidative damage.

As a potent antioxidant, melatonin plays a decisive role in fighting free-radical-related diseases—from cardiovascular disease to cancer and practically everything in between.

In post-menopausal women, for example, melatonin has been found to inhibit *lipid peroxidation* (damage to your fat cells caused by free radicals), thus leading to decreased levels of low-density lipoprotein (LDL) cholesterol, one of the primary ingredients for the formation of atherosclerosis. A newer study on men confirmed these findings, suggesting that melatonin leads to overall lower levels of oxidative stress in humans. In individuals undergoing cardiopulmonary bypass surgery, melatonin exhibited a higher reduction in lipid peroxidation and improvements in red blood cell membrane stiffness. Other widely feared free radical diseases, such as age-related macular degeneration (AMD), acute respiratory distress syndrome (ARDS), glaucoma, and sepsis, have also been responsive to increased melatonin levels.

Melatonin Fights Back Against Cardiovascular Disease

Since cardiovascular disease is the leading cause of death in the United States, melatonin's ability to protect against heart damage is especially noteworthy.

In the past decade, melatonin has received considerable attention, investigating its potential as a cardioprotective nutrient.

Animal studies have provided ample evidence supporting melatonin's antioxidant protection against heart muscle injury, reducing the damage done by a heart attack, and improving the strength of the heart's pumping action following a heart attack.

Other investigators reported that it decreases total cholesterol and LDL levels and increases HDL cholesterol levels. Scientists have discovered that individuals with metabolic syndrome have a lower melatonin production rate compared to healthier counterparts without metabolic syndrome.

Melatonin Fights Back with Cancer and Immune Regulator

Emerging research suggests that melatonin has *anticarcinogenic* properties—that is, it has the ability to prevent cancer from occurring, or to induce cancer cell death if it does occur. This has been attributed to melatonin's antioxidant, anti-inflammatory, anti-proliferative, and hormone-modulating properties.

Melatonin's ability to interfere with cancer cell multiplication and growth ("proliferation"), as well as inducing cancer cell death ("apoptosis"), has been documented in cancer patients.

Melatonin's anticarcinogenic properties can also be attributed to its effect on your immune system. Laboratory studies revealed that melatonin can activate *T-helper cells*, which triggers other immune cells to help kill off foreign invaders or pathogens. Additionally, melatonin stimulates natural killer cell, monocyte, and macrophage synthesis. It has been found to facilitate healthy cell-to-cell communication, which enhances the body's appropriate immune system response to foreign invaders.

Melatonin Protects Against Diabetic Complications

Diabetes—as with cardiovascular disease and cancer—belongs to the family of "free radical diseases." Research has found that people with type 2 diabetes and retinopathy experience alterations in their melatonin secretion.

Considering the large body of evidence identifying melatonin as a significant free-radical scavenger, it is not surprising that preclinical research repeatedly and consistently documents its beneficial antioxidative effects in people with diabetes and those with high blood sugar (*hyperglycemia*).

Melatonin has also been shown to protect pancreatic beta cells and several diabetes-affected organs (including the kidney, retina, brain, and vasculature) from free radical damage. In studies, melatonin treatment has reduced blood glucose, hemoglobin A1c, and cholesterol.

Melatonin Protects Against Obesity with Melatonin

In recent years, dietitians and medical experts have recognized that obesity is often associated with stress, emotional eating, sleep deprivation, and hormonal changes later in life. The study found that women suffering from this disorder had pronounced circadian melatonin rhythm disturbances, which also affected levels of *cortisol* (a stress hormone that can be a factor in weight problems) and *ghrelin* (a hormone that stimulates hunger). It also affected a variety of other behavioral and physiological circadian markers involved in appetite and neuroendocrine regulation.

While no human weight-management trials using melatonin have been published thus far, preclinical trials are encouraging. In middle-aged rodents, daily melatonin administration suppresses abdominal fat, plasma leptin levels, and insulin levels while reducing body weight and food intake. Other researchers reported that melatonin was associated with decreased intra-abdominal fat, decreased plasma insulin and leptin levels, and the absence of age-related weight gain.

Furthermore, laboratory investigations discovered that melatonin can activate brown adipose tissue, which encourages the body to burn fat instead of store it.

Melatonin and Blood Brain Barrier

Melatonin, a naturally occurring free radical scavenger and an inducer of antioxidant enzymes, has been well documented in thousands of publications in the last decade. Melatonin is no longer exclusively classified as a neurohormone since it has been identified in bacteria, fungi, algae, and plants. Likewise, endogenously produced melatonin is no longer the only source in the body since melatonin is also derived from the diet when vegetables, fruits, cereals, herbs, olive oil, wine, or beer are consumed. One important characteristic of melatonin is its permeability into the brain. It readily passes through the blood-brain barrier and accumulates in the central nervous system at substantially higher levels than exist in the blood. As a result, this molecule exhibits neuroprotective solid effects, especially under elevated oxidative stress or intensive neural inflammation.

Summary

"Aging" is defined as the set of gradual and progressive changes in an organism that leads to an increased risk of weakness, disease, and death. This process may occur at the cellular and organ level, as well as in the entire organism of any living being. During aging, there is a decrease in biological functions and the ability to adapt to metabolic stress. General effects of aging include mitochondrial, cellular, and organic dysfunction, immune impairment or inflammation, oxidative stress, and cognitive and cardiovascular alterations, among others. Therefore, one of the main harmful consequences of aging is the development and progression of multiple diseases related to these processes, especially at the cardiovascular and central nervous system levels. Both cardiovascular and neurodegenerative pathologies are highly disabling and, in many cases, lethal. In this context, melatonin, an endogenous compound naturally synthesized not only by the pineal gland but also by many cell types, may have a key role in the modulation of multiple mechanisms associated with aging. This indoleamine is also a therapeutic agent, which may be administered exogenously with a high degree of safety.

Melatonin is best known for its role in the sleep-wake cycle, but it's been studied for its potential pro-aging properties in recent years. Melatonin is a hormone your body naturally makes to help keep your sleep cycle consistent. This cycle is also known as your circadian rhythm. This is also sometimes called the "biological clock." Melatonin plays a significant role in maintaining your sleep cycle. Your body produces most of it in the hours after the sun goes down.

While more research is needed, melatonin is a powerful antioxidant that can help regulate inflammation, the immune system, and mitochondrial function.

- Melatonin *(N-acetyl-5-methoxytryptamine)* is a derivative of the amino acid tryptophan and is widely distributed in food sources, such as milk, almonds, bananas, beets, cucumbers, mustard, and tomatoes.

- In humans, melatonin is primarily synthesized by the pineal gland but is also produced in the gastrointestinal tract and retina.

- Melatonin and its metabolites are potent antioxidants with anti-inflammatory, hypotensive, cell communication-enhancing, cancer-fighting, brown fat-activating, and blood-lipid-lowering effects, thereby protecting tissues from a variety of insults.

- Melatonin has been shown to support circadian rhythm, hormone balance, reproductive health, cognition, mood, blood sugar regulation, and bone metabolism while improving overall antioxidant status and lowering blood pressure.

- Melatonin may help prevent diabetic complications and improve treatment outcomes in patients with cardiovascular disease and certain types of cancer.

Please remember that it's a hormone, so be sure to reach out to a healthcare professional to know whether you can safely take it.

Ref From –

https://www.healthline.com/health/melatonin-and-alcohol#side-effects-of-melatonin

https://www.ncbi.nlm.nih.gov/pmc/articles/PMC9204094/

https://www.lifeextension.com/magazine/2012/9/7-ways-melatonin-attacks-aging-factors?srsltid=AfmBOopZwj_hGnkgVJjBPkteZ9anCNjjXDRA6PmRs9Zxmfid1_HIKaG1

https://www.ncbi.nlm.nih.gov/pmc/articles/PMC3001209/#:~:text=One%20important%20characteristic%20of%20melatonin,than%20exist%20in%20the%20blood.

Martín Giménez, V., De las Heras, N., Lahera, V., Tresguerres, J., Reiter, R., & Manucha, W. (2022). Melatonin as an Anti-Aging Therapy for Age-Related Cardiovascular and Neurodegenerative Diseases. Frontiers in Aging Neuroscience, (), n/a.

Melatonin for Anti-Aging: Is It Effective?. https://www.healthline.com/health/skin/melatonin-anti-aging?utm_source=ReadNext

Beyond Sleep: 10 Ways Melatonin Helps You Age Better (Watch) - GoH. https://garmaonhealth.com/beyond-sleep-10-ways-melatonin-helps-you-age-better-watch/

Melatonin: Anti-Aging and Antioxidant Benefits - HelpHerself.com. https://www.helpherself.com/melatonin-anti-aging-antioxidant-benefits/

Giménez, V. M. M., Heras, N. D. L., Lahera, V., Tresguerres, J. A., Reiter, R. J., & Manucha, W. (2022). Melatonin as an Anti-Aging Therapy for Age-Related Cardiovascular and Neurodegenerative Diseases. Frontiers in Aging Neuroscience. https://doi.org/10.3389/fnagi.2022.888292

Melatonin for Anti-Aging: Is It Effective?. https://www.healthline.com/health/skin/melatonin-anti-aging?utm_source=ReadNext

Chapter – 25

Urolithin - A and Anti-Aging

The aging of an organism is hallmarked by cellular senescence and the accumulation of cellular dysfunction, resulting in the systemic deterioration of tissue. The loss of functional tissue results in increased fragility and can lead to neurodegenerative, musculoskeletal, cardiovascular, integumentary, and neoplastic age-related diseases.

Growing scientific evidence points to mitochondrial dysfunction as a critical contributor to the aging process and the subsequent development of age-related diseases. Senescent tissues, characteristically seen in aging, display various structural and functional changes within mitochondria. Among these observed changes, the accumulation of mitochondrial deoxyribonucleic acid (DNA) mutations and disruptions of oxidative phosphorylation are of particular significance. Furthermore, mitochondrial dysfunction can both cause and result from increased oxidative stress. The mitochondria house a complex collection of energy-producing reactions within the cell, which comprise the process known as mitochondrial respiration. During these cellular energy-producing reactions, the production of chemical agents known as reactive oxygen species (ROS) also occurs.

ROS can cause damage to cellular structures, and thus, an excess of ROS can wreak havoc on the cell. When a cell becomes senescent, the mitochondrial membrane potential is altered, leading to the loss of the proton gradient. This results in an imbalance of energy supply and demand, as well as increased formation of ROS.

These elevated levels of ROS lead to local tissue damage via inflammatory pathways and contribute to neuronal necrosis, seen in neurodegenerative disease, and the accumulation of abnormal mitochondria, seen in sarcopenia.

Moreover, increased ROS can cause further oxidative damage to the mitochondria, resulting in the downregulation of mitochondrial gene expression and further hindrance of oxidative phosphorylation.

Mitophagy and urolithin A

Under normal physiologic conditions, the body removes dysfunctional mitochondria via an autophagic process called mitophagy. Impaired mitophagy can lead to the accumulation of dysfunctional mitochondria, contributing to the development of age-related pathology.

Current research aims to identify potential interventions that promote mitochondrial health and induction of mitophagy in aging tissues.

Urolithin A (UA), a metabolite produced when gut microflora digests the polyphenol compounds ellagitannin and ellagic acid, is a known inducer of mitophagy via several identified mechanisms of action.

Researchers have identified UA as a potent activator of the phosphatase and tensin homolog (PTEN)-induced kinase-1 (PINK1)/parkin-dependent mitophagy pathway, which leads to the selective ubiquitination of mitochondrial proteins for phagolysosomal clearance.

Mitochondrial dysfunction is a hallmark of aging. This derives from reduced production of functional organelles and an accumulation of faulty mitochondria. The consequences are insufficient energy production to support cell functions and a buildup of toxic byproducts detrimental to cellular health.

A promising strategy to reverse mitochondrial dysfunction is to enhance mitophagy, the cellular process to remove and recycle damaged mitochondria. This clears the cells of dysfunctional organelles and, in turn, instructs cells to regenerate healthy organelles.

Urolithin A has been shown to improve mitophagy and mitochondrial function in several tissues affected by aging and age-associated diseases, particularly skeletal muscle. Preclinically, UA enhanced mitochondrial quality in rodent models of aging and muscle dystrophy.

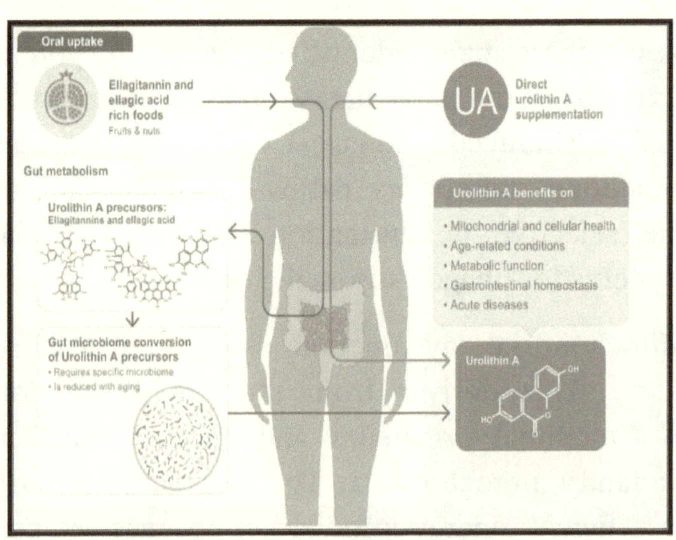

Picture Courtesy From - https://www.cell.com/ trends/molecular-medicine/fulltext/S1471- 4914%2821%2900118-0

Urolithin A is a fascinating compound produced in the human gut from **transforming ellagitannins** in certain fruits like pomegranates, strawberries, and walnuts. However, not everyone's gut bacteria can efficiently convert these precursors into Urolithin A, making supplementation a much more reliable source of Urolithin A for cellular health.

Research on Urolithin A is still ongoing, but there are some interesting fields worth exploring, including:

1. Cellular Mitophagy

2. Skeletal Muscle & Endurance

3. Gut Health

4. Antioxidant Potential

1. Supports Cellular Mitophagy

Mitophagy is a **cellular cleanup process** that removes damaged mitochondria from the body. This is important because mitochondrial dysfunction is linked with molecular damage and leads to age-related diseases.

Think of Urolithin A as a specialized inspector checking the condition of the power plants (ATP energy in mitochondria). In more scientific terms,

Urolithin A activates the process of mitophagy, signaling the body to remove these inefficient or damaged mitochondria.

2. Supports Skeletal Muscle Function and Endurance

Mitochondrial health, muscle function, and endurance go hand in hand, which is why there's a lot of buzz around Urolithin A's potential to help athletes perform at their best and slow down the aging of skeletal muscles.

The mitochondria in our cells are like tiny powerhouses, generating the energy our muscles need to function. As we age, these mitochondria can become less efficient.

Urolithin A promotes mitophagy, helping remove less efficient mitochondria and replace them with newer, more efficient ones. This renewal process ensures that muscle cells have a reliable energy source, which is crucial for maintaining muscle strength and endurance.

With better-performing mitochondria, muscle cells can operate more efficiently. This means they can recover faster from exertion, produce energy more effectively, and ultimately perform better. This is particularly important for aging individuals as their muscle cells naturally lose efficiency over time.

3. Supports Gut Health

Emerging research suggests that Urolithin A may be beneficial in maintaining gut health, a cornerstone of overall wellness and robust immune function. By promoting beneficial bacteria and maintaining a balanced microbiota, Urolithin A could contribute to overall gut health.

How does it do this?

Urolithin A has shown anti-inflammatory properties in some studies. Since inflammation can be detrimental to gut health, leading to conditions like inflammatory bowel disease, Urolithin A's anti-inflammatory effects might be beneficial in maintaining or restoring gut health.

4. Antioxidant Effects

Antioxidants are substances that reduce oxidative stress in the body. Urolithin A has shown antioxidant properties, which might help protect cells from damage caused by free radicals.

Urolithin A helps balance this equation by neutralizing free radicals. It acts like a shield, protecting cells from the harmful effects of these unstable molecules. By doing so, Urolithin A helps reduce oxidative stress, potentially preventing or slowing down cell damage.

Oxidative stress is also a critical factor in the aging process, contributing to cellular damage, inflammation, and the decline of organ function. Urolithin A's antioxidant properties might contribute to healthier aging by protecting cells from age-related damage and chronic inflammation.

Ref From –

Reviewing Evidence for Urolithin A Supplementation – Fight Aging!. https://www.fightaging.org/archives/2023/09/reviewing-evidence-for-urolithin-a-supplementation/?nc

Urolithin A Benefits | Effects, Mitochondrial Function & More. https://neuroganhealth.com/blogs/news/urolithin-a-benefits

https://www.ncbi.nlm.nih.gov/pmc/articles/PMC10460156/

https://www.ncbi.nlm.nih.gov/pmc/articles/PMC10085614/#:~:text=Urolithin%20A%20has%20been%20shown,and%20muscle%20dystrophy%20%5B5%5D.

BY - Katrina Lubiano, based in Canada, Katrina is an experienced content writer and editor specializing in health and wellness. With a journalistic approach, she's crafted over 900,000 words on supplements.

Resources:

Andreux, P. A., Blanco-Bose, W., Ryu, D., Burdet, F., Ibberson, M., Aebischer, P., ... & Rinsch, C. (2019). The mitophagy activator urolithin A is safe and induces a molecular signature of improved mitochondrial and cellular health in humans. Nature Metabolism, 1(6), 595-603.

Liu, S., D'Amico, D., Shankland, E., Bhayana, S., Garcia, J. M., Aebischer, P., ... & Marcinek, D. J. (2022). Effect of urolithin A supplementation on muscle endurance and mitochondrial health in older adults: a randomized clinical trial. JAMA network open, 5(1), e2144279-e2144279.

Singh, A., D'Amico, D., Andreux, P. A., Fouassier, A. M., Blanco-Bose, W., Evans, M., ... & Rinsch, C. (2022). Urolithin A improves muscle strength, exercise performance, and biomarkers of mitochondrial health in a randomized trial in middle-aged adults. Cell Reports Medicine, 3(5).

Kujawska, M., & Jodynis-Liebert, J. (2020). Potential of the ellagic acid-derived gut microbiota metabolite–Urolithin A in gastrointestinal protection. World Journal of Gastroenterology, 26(23), 3170.

Abdelazeem, K. N., Kalo, M. Z., Beer-Hammer, S., & Lang, F. (2021). The gut microbiota metabolite urolithin A inhibits NF-κB activation in LPS stimulated BMDMs. Scientific reports, 11(1), 7117.

D'Amico, D., Andreux, P. A., Valdés, P., Singh, A., Rinsch, C., & Auwerx, J. (2021). Impact of the natural compound urolithin A on health, disease, and aging. Trends in molecular medicine, 27(7), 687-699.

Chapter – 26

Selenium and Anti-aging

Aging is significantly dependent on many metabolic disorders. Since there is a steady trend toward the aging of the human population on a global scale, the number of diseases associated with aging is also gradually increasing. The pathogenesis of various health disorders, including neurodegenerative diseases or cancer, is due to the accumulation of reactive oxygen species (ROS) that cause oxidative stress and inflammation, which are the main contributors to cellular senescence. Generally, aging is characterized by an imbalance between damage inflicted by ROS and the organism's antioxidant defenses. Free radicals could be produced due to the influence of environmental pollutants, metal ions, radiation, or by-products of metabolized drugs. ROS production and oxidative damage to biomacromolecules (nucleic acids, lipids, and proteins) can represent a suitable environment for developing age-related diseases.

Micronutrient deficiencies, such as vitamins and minerals, can suppress immunity and cause a predisposition to infectious diseases, cancer, neurodegeneration, cardiovascular disorders, and hormonal misbalance. Natural antioxidants, as effective free radical scavengers, are indispensable in preventing and treating many age-related disorders.

The trace element **selenium (Se)** was regarded as a dietary supplement for improving health as it possesses valuable antioxidant properties. It may remodel gradual and spontaneous biochemical and physiological changes, potentially leading to disease prevention and healthy aging because it improves antioxidant defense, immune functions, and metabolic homeostasis. **Se** is naturally found in water, soil, and food. Amino acids, peptides, and enzymes are regarded as the main biologically important Se-containing organic compounds.

Se deficiency, which affects about one billion people worldwide, may significantly affect health. In many countries, **Se** may be deficient because of its low soil concentration and, accordingly, in the plants that grow on this substrate.

The risk of Se deficiency increases in proportion to age and age-related diseases. **Se** can be considered a longevity indicator in the elderly population since it plays a role in maintaining aging individuals' health. Inadequate **Se** status may reduce human life expectancy by accelerating aging or increasing vulnerability to various age-related diseases.

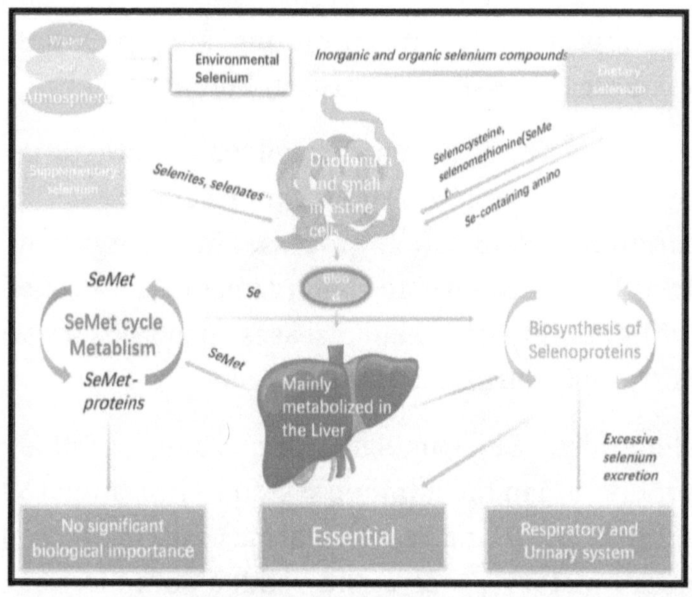

Picture Courtesy From - https://www.researchgate.net/figure/The-process-of-selenium-metabolism-in-human-body_fig3_373915788

In the inevitable aging process provided by nature, there is an imbalance between antioxidative defense and ROS, irreversible changes in mitochondrial renewal, and stem cell exhaustion. According to Alehagen et al., these disorders are closely associated with chronic inflammation, which accompanies age-related diseases. Simonoff et al. claimed that the antioxidant status of older people could be evaluated by measuring **blood Se and vitamin (A and E) levels**. Serum and plasma Se levels, glutathione peroxidase activity, and selenoprotein P concentrations are commonly used measures of a **Se** status in humans.

As a cofactor of enzymes involved in antioxidant protection, Se plays a significant role in regulating different inflammatory processes in the organism. Insufficient **Se** level in the organism is associated with inflammatory skin diseases such as psoriasis and atopic dermatitis.

Generally, **Se** stimulates increasing antibody production in the immune system. The optimal **Se** status (60–175 ng Se/mL plasma) can mitigate an inflammatory process and reduce complications in the lungs, intestines, etc.

As it is known, the proper functioning of the thyroid gland requires several elements, including **Se, Zn, and copper (Cu),** in addition to iodine.

Persistent Se deficiency may also cause infertility. Many clinical studies implicate Se deficiency in several reproductive complications, such as male and female infertility, miscarriage, preterm labor, etc.

Age-related disorders such as neurodegeneration, cardiovascular disease, immune dysfunction, formation of skin wrinkles, etc., are closely associated with **Se** deficiency.

Plants can absorb inorganic **Se** from soil and transform it into an organic form such as selenomethionine or selenocysteine, which are much more accessible for animals and humans than inorganic ones. Consumed by humans, organic **Se** changes by joining amino acids and proteins.

Organic **Se** from food is considered a safe and efficient source of supporting human health. The main animal sources of Se are red meats, poultry, beef or sheep liver, seafood, eggs, and dairy products.

Conclusions

The oxidative damage to macromolecules in the human body represents a suitable environment for developing age-related diseases. The trace element **Se** in the form of selenoproteins improves antioxidant defense, immune functions, and metabolic homeostasis. **Se deficiency** affects about one billion people in the world and may have a significant adverse effect on human health.

Se-containing organic compounds play a key role in the health maintenance of aging individuals. The nutritional doses of **Se** can efficiently stimulate the immune system against infectious diseases or cancer.

Insufficient dietary intake of **Se** can cause cognitive dysfunctions and heart failure in aged persons. Organic **Se** from food of animal, plant, or mushroom origin is considered a safe and efficient source of supporting human health.

Ref From –

https://www.ncbi.nlm.nih.gov/pmc/articles/
PMC9570904/

Molecules | Free Full-Text | Selenium: An Antioxidant with a Critical Role in Anti-Aging. https://www.mdpi.com/1420-3049/27/19/6613

Chapter – 27

Vitamin D3 and Anti-aging

Vitamin D synthesis. Main forms of vitamin D in nature are: vitamin D2 (ergocalciferol) that is photochemically synthesized in plants, and vitamin D3 (cholecalciferol) that is synthesized in the skin of animals and humans in response to sunlight. The synthetic pathway involves 25- and 1-alpha-hydroxylation of vitamin D2 and D3, in the liver and kidney, respectively. First hydroxylation occurs within the liver and lead to the formation of 25(OH)D or calcidiol; second hydroxylation occurs within the kidneys and constitutes the most biologically active hormonal form of vitamin D: 1,25(OH)2D, or calcitriol.

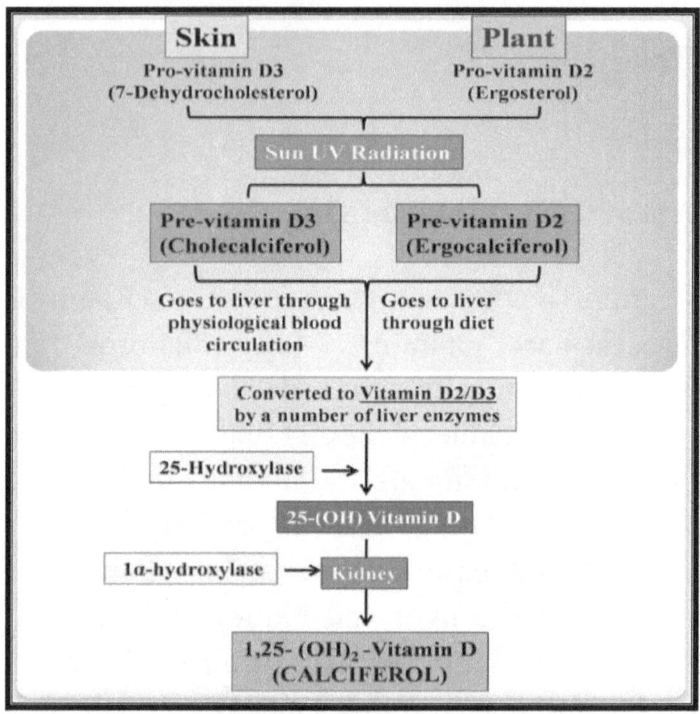

Vitamin D, also known as calciferol, comprises a group of fat-soluble seco-sterols. The two major forms are vitamin D_2 and vitamin D_3. Vitamin D_2 (ergocalciferol) is largely human-made and added to foods, whereas vitamin D_3 (cholecalciferol) is synthesized in the skin of humans from 7-dehydrocholesterol and is also consumed in the diet via the intake of animal-based foods. Both vitamin D_3 and vitamin D_2 are synthesized commercially and found in dietary supplements or fortified foods. The D_2 and D_3 forms differ only in their side chain structure. The differences do not affect metabolism (i.e., activation), and both forms function as prohormones.

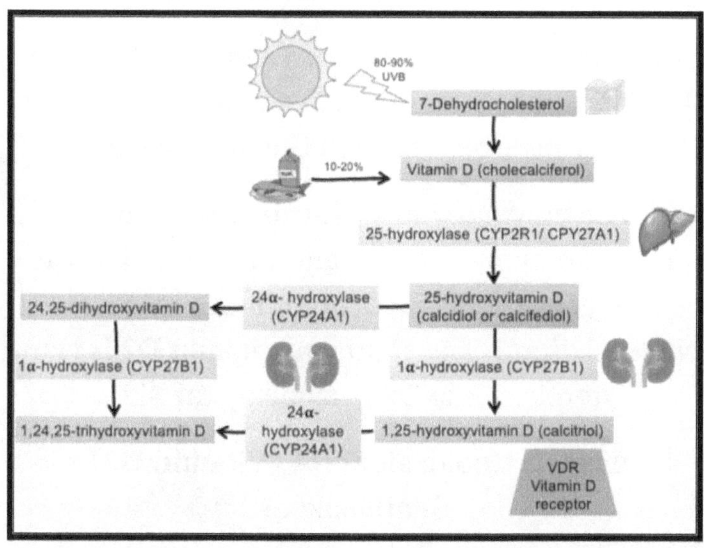

Picture Courtesy From - https://www.researchgate. net/figure/The-vitamin-D-endocrine-system-metabolism_fig1_361295995

Vitamin D3 is crucial in regulating physiological processes such as calcium metabolism (for bone development and health), immunity, cellular growth, and cardiovascular function. Vitamin D3 or cholecalciferol is crucial for bone health, boosts your immune system, and can help improve cardiovascular functioning.

Given the pivotal role vitamin D plays in the body, insufficient quantities could hamper one's level of activity and productivity and ultimately compromise one's health.

Choleciferol is one form of vitamin D. Vitamin D, along with vitamins A, E, and K, is fat-soluble and is absorbed through the digestive tract.

The only nutritional forms of vitamin D found in humans are vitamin D2, Also known as ergocalciferol, and vitamin D3, known as cholecalciferol. The source of vitamin D2 is plant sterols, while D3 is made by the skin.

How is Cholecalciferol (Vitamin D3) Synthesized?

Vitamin D3 is manufactured from a particular lipid (a zoosterol) called 7-dehydrocholesterol (7-DHC), found in the skin's epidermis layer. This lipid is also involved in the biosynthesis of cholesterol.

When exposed to solar radiation, the sterol absorbs photons from the ultraviolet B (UVB) radiation range. This results in a photochemical reaction known as photolysis, which forms cholecalciferol or pre-vitamin D3. It then undergoes another process called thermal isomerization, which reorganizes it into a more stable inactive vitamin D3 or cholecalciferol molecule.

When newly synthesized, vitamin D3 is inactive and has low solubility in water. Therefore, it moves out of the plasma membrane of skin cells into the extracellular fluid. It attaches itself to the vitamin D3 binding protein (DBP) and is transported through the blood to the liver.

In the liver, the inactive vitamin D3 undergoes a process called hydroxylation to convert it to calcidiol 25-hydroxyvitamin D (25(OH)D). This molecule is then transported to the kidney, where it's converted from calcidiol to calcitriol or 1,25(OH)D.

While calcitriol is the most active form of vitamin D3, calcidiol is normally used to indicatef vitamin D3 status in a person. So when you get a blood test, this will be the form of vitamin D that is measured and it's the most abundant form of vitamin D3 in the circulatory system.

It's important to note that healthy liver and kidney function plays a vital role in converting the inactive cholecalciferol (vitamin D3) into the more active forms of calcidiol and calcitriol.

Vitamin D3 helps calcium and phosphorous absorption - essential minerals for bone health

Optimal absorption of calcium and phosphorus is crucial in preventing osteoporosis and protecting bone. The total amount of calcium the body absorbs from the diet is dependent on two main factors: the actual quantity consumed and the efficiency of the absorption process, which is regulated by vitamin D3.

In a 2008 research study, researchers showed that vitamin D3 deficiency ultimately causes calcium malabsorption. In the intestine, calcitriol, the active form of vitamin D3, binds to vitamin D receptors (VDR) and stimulates the calcium transport system.

For phosphorus absorption, calcitriol increases the levels of the sodium-dependent phosphate transport system, which helps the cells absorb more phosphorus.

Vitamin D3 May Help Protect the Brain and Preserve Neurological Development

A 2012 study showed that calcitriol (the active form of vitamin D 3) has some neuroprotective effects. For instance, it was found to be effective in eliminating amyloid plaques, a typical characteristic of Alzheimer's.

In an in-depth literature review of the role of vitamin D3 in neurological development and brain health, the researchers concluded that mounting evidence suggests that vitamin D is required for normal brain development and function. They found an association between low levels of vitamin D3 and a broad range of neurological conditions such as Alzheimer's Disease, Parkinson's Disease, multiple sclerosis, and other neurocognitive disorders. The researchers concluded that more studies are needed in this field, as vitamin D supplementation is widely available.

Vitamin D3 May Help Fracture Repair

There are two ways to examine the relationship between vitamin D3 and fractures. First, when vitamin D3 levels are low, individuals may develop conditions such as osteoporosis and weak muscles, increasing the risk of falls and fractures.

When fractures occur, vitamin D3 helps in the healing process. In a small study on vitamin D status and adult fracture healing, researchers found that fracture healing was delayed in participants who were vitamin D deficient. They concluded that the research seems to indicate that vitamin D status at the time of fracture affects fracture healing.

Vitamin D3 May Help Cardiovascular Health

Arterial hypertension is one of the main risk factors behind global morbidity. Vitamin D3 and its effect on cardiovascular health have attracted extensive scientific research, although the results have been inconsistent, and more studies are needed. In a literature review, researchers concluded that the current evidence suggests that lower vitamin D levels correlate with a higher risk of cardiovascular disease and related conditions, but more conclusive research is needed.

Vitamin D3 May Help Strengthen the Immune System

The vitamin D receptor (VDR) is present in immune cells such as antigen-presenting T and B cells. This means that the cells can synthesize the active form of vitamin D3. Research has shown that vitamin D3 can regulate your immune response, hence boosting autoimmunity and minimizing susceptibility to infections.

Ref From –

https://www.ncbi.nlm.nih.gov/books/
NBK56061/#:~:text=Vitamin%20D3%20is%20
synthesized,wavelength%20290%20to%20
320%20nm.

https://www.researchgate.net/figure/Vitamin-
D-synthesis-Main-forms-of-vitamin-D-in-nature-
are-vitamin-D2-ergocalciferol_fig1_225079618

https://www.researchgate.net/figure/A-
schematic-representation-of-the-synthesis-
of-active-forms-of-vitamin-D-from-the-skin_
fig3_344450402

https://youthandearth.com/blogs/blog/vitamin-
d3-cholecalciferol-benefits

Chapter – 28

Magnesium and Anti-aging

Aging is both universal and inevitable. During aging, changes occur at biological, psychological, and physiological levels. Some of these changes are benign, such as greying hair. Others result in the declining function of senses and activities of daily living and increased susceptibility to disease, frailty, and disability. In fact, advancing age is the significant risk factor for numerous chronic diseases.

In 2013, Lopez-Otin et al. proposed nine hallmarks of aging, comprising genomic instability, telomere attrition, epigenetic alterations, mitochondrial dysfunction, loss of proteostasis, deregulated nutrient sensing, cellular senescence, stem cell exhaustion, and altered intercellular communication. Indeed, new hallmarks of aging have been added to the original ones, including autophagy, microbiome disturbance, and inflammation, among other emerging ones. Advances in understanding the mechanisms underlying the aging process and the possible modifiable determinants can reveal insights on achieving the healthiest possible aging. Magnesium is a fundamental mineral, indispensable for numerous cellular processes including all oxidative phosphorylation processes, over 600 enzymatic reactions, energy generation, nucleic acids synthesis and stability, protein synthesis, and carbohydrate metabolism. Adequate magnesium status is essential for organs and systems in the human body. Magnesium deficiency is common in late life and has been associated with various age-related chronic diseases.

Magnesium is essential for the active transport of other ions across cell membranes; it modulates muscle contraction, normal cardiac rhythm, and neuron excitability. Magnesium is involved in adenosine triphosphate (ATP) mitochondrial synthesis to form MgATP, which is necessary for crucial cellular reactions and signaling, including all protein phosphorylation reactions and cyclic adenosine monophosphate (cAMP) activation, which is involved in numerous biochemical cellular processes, comprising ribonucleic acid (RNA) expression, deoxyribonucleic acid (DNA) synthesis, glucose metabolism, muscular and neural cell signaling, and blood pressure control.

Chronic magnesium deficiency is frequent among older adults, which can be explained by various reasons, including a low dietary magnesium content (usual in Western diets), increased urinary excretion, or reduced intestinal absorption owing to several pathological conditions.

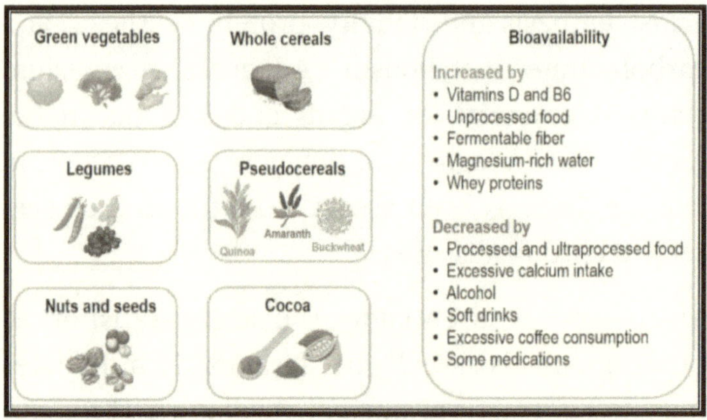

Picture Courtesy From - https://www.ncbi.nlm.nih.gov/pmc/articles/PMC10892939/

Diseases that can be associated with hypomagnesemia

Diabetes

Hypertension

Malnutrition (inadequate intake)

Alcohol abuse

Chronic diarrhea

Hyperthyroidism

Mild deficits of magnesium are usually asymptomatic, and, when apparent, clinical signs are typically absent or non-specific and may be confused with common symptoms associated with aging.

Symptoms and signs of hypomagnesemia

Neuromuscular and central nervous system

Muscle cramps

Muscle weakness, fasciculations, tremors

Lethargy, tetany

Vertigo, nystagmus

Depression, psychosis

Carpopedal spasm

Convulsions

Electrolyte disturbance

Hypokalemia

Hypocalcemia

Other disorders

Alterations in glucose homeostasis

Hypertension

Atherosclerotic vascular disease

Myocardial infarction

Osteoporosis

Asthma

Migraine

Chronic fatigue syndrome

Hallmarks of aging, which have been closely related to magnesium

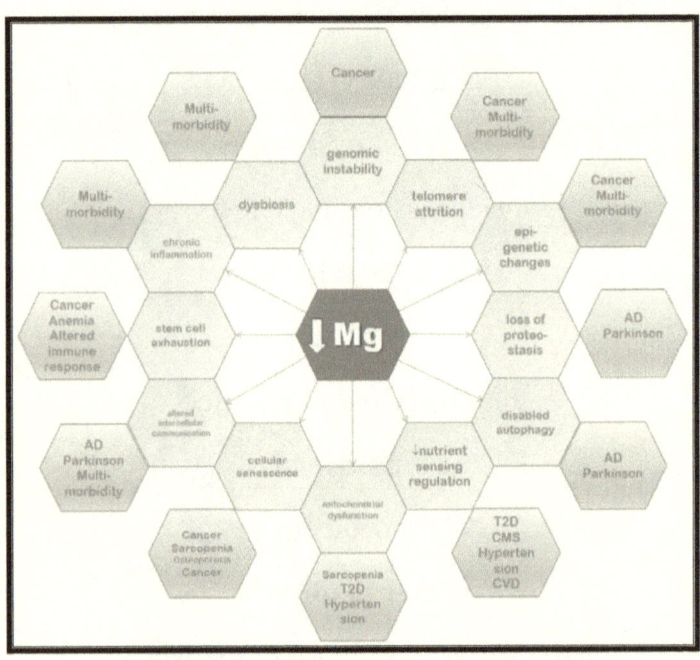

Picture Courtesy From - https://www.ncbi.nlm.nih.gov/pmc/articles/PMC10892939/

Low magnesium status is associated with all the hallmarks of aging (Inside the circle). Examples of age-related diseases connected with the hallmarks of aging are shown (outside the circle). AD: Alzheimer's disease; CMS: cardiometabolic syndrome; CVD: cardiovascular disease; and T2D: type 2 diabetes.

Genomic instability refers to an increased predisposition to genomic modifications (e.g., DNA damage, mutations, and chromosomal abnormalities), and the effects of epigenetic alterations, oxidative stress, and deficient DNA repair and telomere maintenance engender it.

Telomeres are sections of repetitive nucleotide sequences containing non-essential information located at both ends of each chromosome. They protect against degradation and fusion with other chromosomes, ensuring that no genetic information is lost and controlling the number of cell replications.

Mitochondria are the cellular 'powerhouse' producing most of the available ATP, which is the main energy source for cellular processes. They have their genome, which is susceptible to damage, being stored in a pro-oxidation location. Mitochondria are the core for multiple signaling cascades driving the cell destiny to either survival or death by apoptosis. They are the main sources of free radicals (reactive oxygen species, ROS) as a consequence of normal cellular metabolism. The accumulation of dysfunctional mitochondria is characteristic of aging, with decreased ATP production and increased ROS generation. Excessive ROS damages all molecules, from proteins to DNA, causing them to mutate and, thereby, dysregulating their function. Dysfunctional mitochondria produce less ATP, reducing the energy supply and leading, over time, to chronic inflammation, oxidative stress, and cellular damage.

Aging is associated with significant alterations in cell-to-cell communication via neuronal, endocrine, and neuroendocrine routes. Inflammation is one of the most extensively studied and relevant intercellular communication events and it is modified with aging. Communication between cells is essential for coordinating cell functioning at all levels (organs, tissues, whole body) and encompasses soluble factors, including cytokines, chemokines, neurotransmitters, and growth factors, recognized by specific receptors at the cell surface.

We have already mentioned that age-related chronic low-grade inflammation, or inflammaging, is involved in a wide range of chronic diseases. Aging correlates with high blood levels of inflammatory mediators, such as IL-1, IL-6, C-reactive protein (CRP), interferon (IFN)α, and several others.

Ref From –

Dominguez, L., Veronese, N., Barbagallo, M., & Barbagallo, M. (2024). Magnesium and the Hallmarks of Aging. Nutrients, 16(4), 496.

https://www.ncbi.nlm.nih.gov/pmc/articles/ PMC10892939/

Chapter – 29

Amino acids and Aging-related diseases

Proteins are made up of 20 amino acids. Each amino acid has an α-carboxyl group, a primary α-amino group, and a side chain called the R group. Amino acids are the fundamental building blocks of proteins and nitrogenous backbones for compounds such as neurotransmitters and hormones. In chemistry, an amino acid is an organic compound containing an amino functional group ($-NH_2$) and a carboxylic acid functional group ($-COOH$).

Amino acids are the building blocks of protein. Proteins are long chains of amino acids. Your body has thousands of different proteins that each have important jobs. Each protein has its own sequence of amino acids. The sequence makes the protein take different shapes and have distinct functions in your body.

You can think of amino acids like the letters of the alphabet. When you combine letters in various ways, you make different words. The same goes for amino acids — when you combine them in various ways, you make different proteins.

Your body makes hundreds of amino acids, but it can't make nine of the amino acids you need. These are called essential amino acids. You must get them from the food you eat. The nine essential amino acids are:

Histidine: Histidine helps make a brain chemical (neurotransmitter) called histamine. Histamine plays an important role in your body's immune, digestion, sleep, and sexual functions.

Isoleucine: Isoleucine is involved with your body's muscle metabolism and immune function. It also helps your body make hemoglobin and regulate energy.

Leucine: Leucine helps your body make protein and growth hormones. It also helps grow and repair muscle tissue, heal wounds, and regulate blood sugar levels.

Lysine: Lysine is involved in the production of hormones and energy. It's also important for calcium and immune function.

Methionine: Methionine helps with your body's tissue growth, metabolism and detoxification.

Methionine also helps with the absorption of essential minerals, including zinc and selenium.

Phenylalanine: Phenylalanine is needed for the production of your brain's chemical messengers, including dopamine, epinephrine and norepinephrine. It's also important for the production of other amino acids.

Threonine: Threonine plays an important role in collagen and elastin. These proteins provide structure to your skin and connective tissue. They also help with forming blood clots, which help prevent bleeding. Threonine plays an important role in fat metabolism and your immune function, too.

Tryptophan: Tryptophan helps maintain your body's correct nitrogen balance. It also helps make a brain chemical (neurotransmitter) called serotonin. Serotonin regulates your mood, appetite and sleep.

Valine: Valine is involved in muscle growth, tissue regeneration and making energy.

Nine amino acids cannot be synthesized by human or other mammalian cells. Therefore, these amino acids must be supplied from an exogenous diet.

Your body produces the rest of the 11 amino acids you need. These are called nonessential amino acids. The nonessential amino acids are alanine, arginine, asparagine, aspartic acid, cysteine, glutamic acid, glutamine, glycine, proline, serine and tyrosine.

Some nonessential amino acids are classified as conditional. This means they're only considered essential when you're ill or stressed. Conditional amino acids include arginine, cysteine, glutamine, tyrosine, glycine, ornithine, proline and serine.

Among these 20 amino acids, 9 are essential—phenylalanine, valine, tryptophan, threonine, isoleucine, methionine, histidine, leucine, and lysine.

Amino acids are molecules that combine to form proteins. Amino acids and proteins are the building blocks of life.

When proteins are digested or broken down, amino acids are the result. The human body then uses amino acids to make proteins to help the body:

- Break down food
- Grow

- Repair body tissue

- Perform many other body functions

Amino acids can also be used as a source of energy by the body.

Amino acids are classified into three groups:

- Essential amino acids

- Nonessential amino acids

- Conditionally essential amino acids

ESSENTIAL AMINO ACIDS

- Essential amino acids cannot be made by the body. As a result, they must come from food.

- The 9 essential amino acids are: histidine, isoleucine, leucine, lysine, methionine, phenylalanine, threonine, tryptophan, and valine.

NONESSENTIAL AMINO ACIDS

Nonessential means that our bodies can produce the amino acid, even if we do not get it from the food we eat. Nonessential amino acids include: alanine, arginine, asparagine, aspartic acid, cysteine, glutamic acid, glutamine, glycine, proline, serine, and tyrosine.

CONDITIONALLY ESSENTIAL AMINO ACIDS

- Conditionally essential amino acids are usually not essential, except in times of illness and stress.

- Conditionally essential amino acids include: arginine, cysteine, glutamine, tyrosine, glycine, proline, and serine.

What do amino acids do?

Your body uses amino acids to make proteins. The different types of amino acids and the way they're put together determine the function of each protein. So, amino acids are involved in many important roles in your body. Amino acids help:

Break down food.

Grow and repair body tissue.

Make hormones and brain chemicals (neurotransmitters).

Provide an energy source.

Maintain healthy skin, hair and nails.

Build muscle.

Boost your immune system.

Sustain a normal digestive system.

It's not surprising that everybody gets old. Aging, a natural and inevitable aspect of life, is defined as a progressive decrease in physiological functions [1]. Skin, an important protective barrier between the body and the external environment, is the largest human organ on which the effects of aging processes are the most visible. Wrinkles on the face are the first signs of the passing of time. Skin aging is a complex biological process influenced by not only proper internal (also known as endogenous or intrinsic) - genetic (including cellular metabolism or hormones) but also improper external (exogenous, extrinsic) - environmental (in this chronic light exposure, pollution, ionizing radiation, chemicals, toxins, non-balanced diet) factors. In other words, this is an "aging mosaic," which results from destructive reactions occurring in different cells. Generally speaking, the structure of the skin is mainly made of collagen and elastin proteins. Collagen is the most important component of the extracellular matrix, determining skin physiology (its structure and functions). The name collagen means glue. It defines its basic function – gluing cells together.

Amino acids are either produced within the body (non-essential amino acids) or can be found in many food sources (essential amino acids) [14].

The proper balance of amino acids level should be maintained due to the regulation of skin condition. Nevertheless, the level of amino acids decreases with age leading to the body's inability to regenerate skin cells in an effective way. On the other hand, a diet containing amino acids (meat, eggs, tuna, eggs, salmon, hemp seeds, quinoa, Goji-red berries, nuts, hummus, tofu, chickpeas, milk or supplements) supports skin health.

All twenty standard amino acids play a role in the regulation of aging, creating and maintaining smooth, healthy, younger-looking skin. They are natural, safe, nontoxic, environmentally friendly, sustainable resources and do not cause allergic reactions. They activate collagen synthesis and have antioxidant properties.

To sum up, amino acids have nutritional benefits and great potential for the treatment of aging and aging-related diseases.

Ref From –

https://www.ncbi.nlm.nih.gov/books/NBK557845/#:~:text=Nutritionally%2C%20amino%20acids%20are%20divided,essential%20during%20periods%20of%20stress.https://medlineplus.gov/ency/article/002222.htm

https://my.clevelandclinic.org/health/
articles/22243-amino-acids

https://austinpublishinggroup.com/nutritional-
sciences/fulltext/ijns-v5-id1039.php

TONG, K., Au Yeung, H., Chan, C., & YU, Z.
(2017). Copper-based reactions in analyte-
responsive fluorescent probes for biological
applications. Journal of Inorganic Biochemistry.
https://doi.org/10.1016/j.jinorgbio.2017.07.001

Amino Acid: Benefits & Food Sources.
https://my.clevelandclinic.org/
health/articles/22243-amino-acids?_
ga=2.232099871.145207631.1678553185-
96067237.1678211649&_gl=1*19kbjlq*_
ga*OTYwNjcyMzcuMTY
3ODIxMTY0OQ..*_ga_
HWJ092SPKP*MTY3ODU2MTAzNC40LjEuMT
Y3ODU2MjQ0MC4wLjAuMA..

https://www.sciencedirect.com/science/article/
pii/S2468501119300082

Chapter – 30

Spermidine and Anti-aging

Spermidine is a polyamine compound implicated in cellular survival, growth, and proliferation and is also known for its neuroprotective, cardioprotective, anti-cancer, and anti-inflammatory properties.

Polyamines (PAs) are a subclass of bioactive amines that perform a variety of functions in living cells. Many of the biogenic (i.e., biologically formed) amines are generated from different amino acids or related metabolites by decarboxylase reactions.

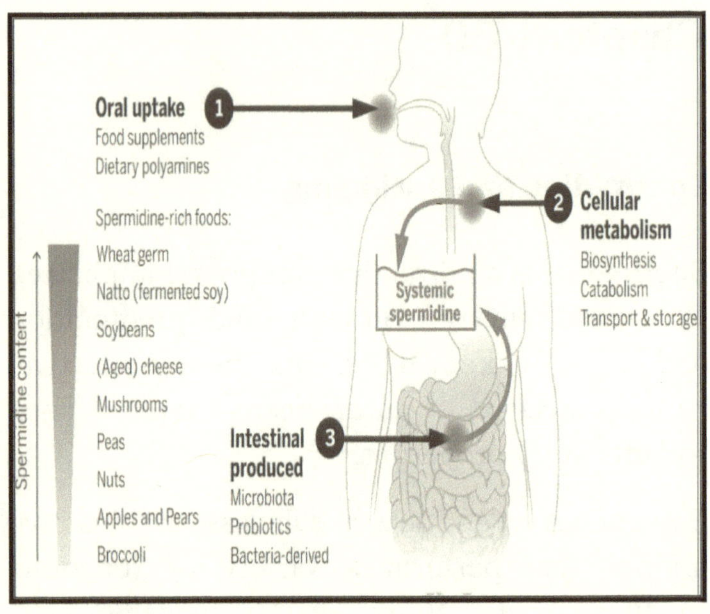

Picture Courtesy From - https://www.science.org/
doi/10.1126/science.aan2788

Scheme depicting the sources crucial for spermidine bioavailability in the whole organism. In addition to cellular metabolism, spermidine is taken up orally from dietary sources or produced by commensal gut bacteria. Subsequently, spermidine can be resorbed by intestinal epithelial cells and is distributed through systemic circulation. Examples of spermidine-rich foods are enumerated.

Food supplements, including the polyamine precursor **arginine** and probiotics (polyamine-producing bacteria), increase the intestinal production of polyamines.

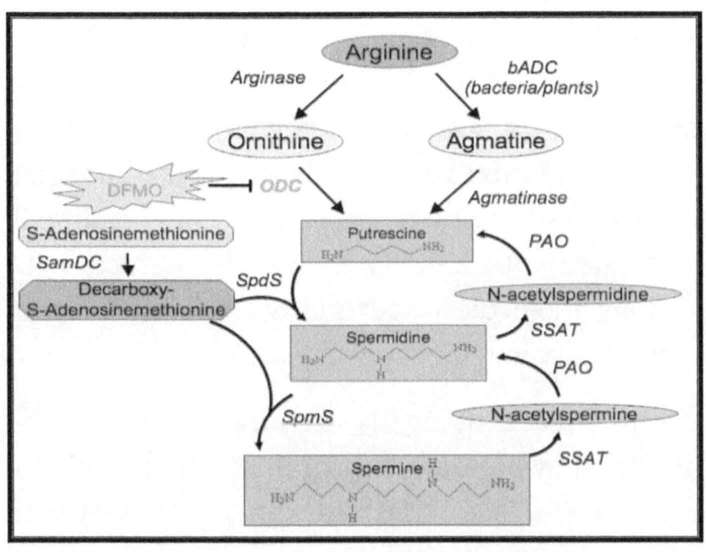

Picture Courtesy From - https://www.weizmann.ac.il/molgen/Kahana/polyamines

Polyamines are ubiquitous polycations found in all cells, tissues, and organs. They can interact with negatively charged molecules such as DNA, RNA, adenosine triphosphate, and proteins. These molecules exert multiple functions in many physiological and pathophysiological processes, including cell proliferation, differentiation, growth, tissue regeneration, and gene regulation.

Due to its antioxidant functions, anti-inflammatory properties, enhanced proteostasis, and improved mitochondrial metabolic functions, spermidine, a naturally occurring polyamine, is involved in a series of biological events, including autophagy induction, apoptosis, transcription, and DNA stability. The spermidine concentration declines with age, and exogenous spermidine supplementation reverses age-associated adverse changes and prolongs the lifespan. Spermidine is associated with longevity. Since it interacts with various molecules, spermidine influences aging through diverse mechanisms.

Even though aging is inevitable, it can be modified by biological and genetic interventions, pharmaceuticals, lifestyle, and the systemic environment. Spermidine has been shown to be important for prolonging survival outcomes, and abnormal changes in spermidine levels are associated with aging and disease development. Intracellular concentrations of spermidine are suppressed during aging. Exogenous spermidine supplementation has been shown to extend the lifespans of flies, nematodes, and yeast. Moreover, a diet enriched in spermidine was shown to prolong the lifespans of mice.

However, studies on the mechanisms of action of spermidine are rare. Autophagy is the primary mechanism of spermidine, which delays aging and prolongs lifespan. In addition, spermidine exerts its effects through other mechanisms, including anti-inflammation, histone acetylation reduction, lipid metabolism, and regulation of cell growth and signaling pathways.

Recent studies have indicated that spermidine levels decline with age in various tissues. Many studies have shown that spermidine administration may promote life span in different model organisms. The main reason for this effect is assumed to be the induction of autophagy and oxidative stress suppression. It was shown that spermidine combined with exercise could restore mitochondrial function in cardiac muscle cells and slow aging. Spermidine supplementation was also found to slow the aging of neuronal cells by restoring mitochondria functions.

Moreover, spermidine activated mitophagy by affecting the mitochondrial membrane integrity, followed by ATM kinase activation and PINK1/ Parkin-dependent mitophagy induction. In addition, a recent study conducted in mice has shown the induction of mitochondrial biogenesis by modulating SIRT1 activity and NRF1, NRF2, and TFAM levels, resulting in improved cardiac functioning.

Ref From –

https://www.aginganddisease.org/
EN/10.14336/AD.2021.0603

https://www.annualreviews.org/content/
journals/10.1146/annurev-nutr-120419-015419

https://www.sciencedirect.com/topics/
chemistry/spermidine

Chapter – 31

Autoimmune diseases and cancer

Autoimmunity is a condition in which the body's immune system fights off its own tissues, causing harmful complications. Abnormalities in T-cell activities and the expression of autoantibodies are the main causes of this phenomenon.

Like the two ends of a magnet, cancer, and autoimmunity share a common origin but exert powerful forces that work in opposite directions. Both diseases result from failures in the body's immune system. Cancer often develops because the immune system fails to attack defective cells, allowing them to divide and grow. Conversely, autoimmunity—a faulty immune response that leads to diseases such as colitis and lupus—occurs when the immune system mistakenly attacks healthy cells.

Can autoimmune disease cause cancer?

Autoimmune disorders generally attack a single organ or part of the body, often causing inflammation in the affected area. While inflammation is not a direct cause of cancer, chronic inflammation may increase cancer risk. Autoimmune diseases affect the gastrointestinal tract - Inflammatory bowel disease, Crohn's disease, and colitis. For instance, it causes chronic inflammation in the digestive system that increases the risk of colorectal cancer. Chronic inflammation may damage cell DNA, leading to uncontrolled cell growth, one of the hallmarks of cancer.

Autoimmune disease may also lead to a higher risk of cancers of the blood, bone marrow, and lymph nodes, such as leukemia and lymphoma.

Certain autoimmune diseases and their treatments increase the risk of getting cancer. And some cancers and cancer treatments increase the risk of autoimmune conditions. It goes both ways.

The immune system is your body's army. It protects you against things that don't belong, like germs and cancers. When the immune system detects cancer, it sounds an alarm and activates the troops to fight it.

At this point, there are three possible outcomes:

- The immune system wins and kills the cancer. You never even know the cancer was there.

- The immune system tries its best, but the cancer grows too fast. The cancer develops.

- The immune system activates a lot of troops, but things go awry. While trying to fight the cancer, the immune system gets confused and starts attacking your body, too. This is what we call autoimmune disease ("auto," meaning "self")

Do autoimmune diseases increase the risk of cancer?

Autoimmune conditions can increase the risk of cancer in many different ways. We don't completely understand why. But scientists are learning more about this complicated relationship every year.

Here are a few things:

- Autoimmune diseases cause chronic inflammation, which can stimulate cancer growth. Inflammation can change the cell's DNA (its instruction manual). This allows cells to reproduce too quickly and form cancers.

- Inflammation can increase blood vessel growth, feeding cancers.

- Certain autoimmune conditions can make it harder for the body to clear viruses that cause cancers. An example is human papillomavirus (HPV), which causes cervical cancer.

Ref From –

https://www.cancercenter.com/risk-factors/autoimmune-diseases#:~:text=Can%20autoimmune%20disease%20cause%20cancer,inflammation%20may%20increase%20cancer%20risk.

https://www.goodrx.com/health-topic/autoimmune/autoimmune-disease-and-cancer

https://www.sciencedirect.com/science/article/abs/pii/B9780323854153000180#:~:text=Abstract,also%20a%20matter%20of%20concern.

Uddin, J., Tran, D. K., Hannan, M. A., H. A. F., Rahman, M. A., Moni, A., Lâm, N. T., Ngọc, V. T. N., & Chu, D. (2022). Autoimmune diseases and metabolic disorders: Molecular connections and potential therapeutic targets. Elsevier EBooks. https://doi.org/10.1016/b978-0-323-85415-3.00018-0

Chapter – 32

Omega Fatty 3 Acid and Anti-aging

Life expectancy is increasing globally, and the prevalence of age- and lifestyle-related non-communicable diseases (NCDs), such as cancer, heart disease, respiratory disease, type 2 diabetes, obesity, chronic kidney disease, and dementia, is rising. This has led patients to present with multiple co-morbidities, creating more complex needs (e.g., the need for multiple medications) and putting significant pressure on healthcare and social systems. Undernutrition and overnutrition can both seriously impact an individual's risk of developing an NCD. There is, therefore, a growing demand for appropriate nutrition interventions and targeted medical nutrition supplements or formulas to address patient needs, improve outcomes, and help reduce healthcare costs. Inflammation is considered to play a central role in age- and lifestyle-related NCDs, in loss of muscle mass and strength (sarcopenia) in frailty and cancer, and in the response to surgery and in critical illness.

Hence, targeting inflammation is thought to be appropriate to disease prevention and treatment. The long-chain omega-3 polyunsaturated fatty acids (LCPUFAs), docosahexaenoic acid (DHA), and eicosapentaenoic acid (EPA) are known to have roles in supporting human health, with one of their primary actions being to reduce inflammation and promote its resolution. This suggests a broad role for DHA and EPA in the prevention and treatment of disease, including, but not restricted to, specific therapeutic areas such as age-related decline in muscle mass, oncology, perioperative care, and cognitive health.

Like all mammals, humans cannot synthesize the essential omega-3 fatty acid α-linolenic acid. Furthermore, endogenous synthesis of EPA and DHA from α-linolenic acid is described as poor in most humans and is influenced by a range of factors such as age, sex, genetics, and disease. Therefore, performed EPA and DHA must be obtained from the diet or supplements.

Blood levels of EPA and DHA are highly related to intake. Global mapping indicated low or even very low blood levels of omega-3 LCPUFAs (i.e., DHA and EPA) in a large proportion of people for whom data were available, suggesting low intakes in those populations. Reliance on the endogenous synthesis of EPA and DHA is challenged by the low activity of this pathway, which is further impaired in conditions such as insulin resistance. Therefore, the benefits of DHA and EPA might be particularly pronounced in those population groups with insulin resistance or other features that limit endogenous synthesis. The anti-inflammatory and inflammation-resolving effects of DHA and EPA have been shown to be relevant to improving clinical outcomes in several specific therapeutic areas.

Omega-3 fatty acids are "healthy fats" that may support your heart health. One key benefit is helping to lower your triglycerides. Specific types of omega-3s include DHA and EPA (found in seafood) and ALA (found in plants). Some foods that can help you add omega-3s to your diet include fatty fish (like salmon and mackerel), flaxseed and chia seeds.

Omega-3 fatty acids (omega-3s) are polyunsaturated fats that perform important bodily functions. **Your body cannot produce the amount of omega-3s you need to survive. So, omega-3 fatty acids are essential nutrients, meaning you need to get them from your food.**

The two main types of fatty acids are saturated fat and unsaturated fat. Unsaturated fat further breaks down into polyunsaturated fat and monounsaturated fat. These are terms you commonly see on nutrition labels.

Fatty acids are chain-like chemical molecules made up of carbon, oxygen and hydrogen atoms. Carbon atoms form the backbone of the chain, with oxygen and hydrogen atoms latching on to available slots.

Saturated fat has no more open slots. A monounsaturated fat has one open slot. A polyunsaturated fat has more than one open slot.

Saturated fats are sometimes known as "bad" or "unhealthy" fats because they increase your risk of certain diseases like heart disease and stroke. Unsaturated fats (polyunsaturated and monounsaturated) are considered "good" or "healthy" fats because they support your heart health when used in moderation.

Omega-3s, as a form of polyunsaturated fat, are healthier alternatives to saturated fat in your diet.

Omega-3 fatty acids help all cells in the body function properly. They are a vital part of cell membranes, helping to provide structure and supporting interactions between cells. While they are essential to all cells, omega-3s are concentrated in high levels in the eyes and brain cells.

In addition, omega-3s provide the body with energy (calories) and support the health of many body systems. These include your cardiovascular system and endocrine system.

There are three main types of omega-3 fatty acids:

EPA (eicosapentaenoic acid). EPA is a "marine omega-3" because it's found in fish.

DHA (docosahexaenoic acid). DHA is also a marine omega-3 found in fish.

ALA (alpha-linolenic acid). ALA is the form of omega-3 found in plants.

Omega-3s are essential nutrients that you need to get from the diet. When you get ALA from food, the body is able to turn some of the ALA into EPA and subsequently to DHA. However, this process provides just a small amount of EPA and DHA. So, dietary sources of EPA and DHA (like fish) are essential.

Omega-3 fatty acids have many potential benefits for cardiovascular health. One key benefit is that they help lower triglyceride levels. Too many triglycerides in the blood (hypertriglyceridemia), raise the risk of atherosclerosis, and through this, can increase your risk of heart disease and stroke. So, it is essential to keep triglyceride levels under control. In addition, omega-3s may help by raising the HDL (good) cholesterol and lowering your blood pressure.

Some studies show omega-3s may lower your risk for:

Cardiovascular disease (CVD).

Blood clots.

Beyond heart health, omega-3s may help lower your risk of developing:

Some forms of cancer, including breast cancer.

Alzheimer's disease and dementia.

Age-related macular degeneration (AMD).

Research continues to investigate these and other possible benefits.

You can look to certain plant-based sources of omega-3, which provide the nutrient in the form of ALA. Alternatively, you can speak with your provider or doctor about supplements.

One of the best sources of ALA is ground or milled flaxseed. Aim to add about 2 tablespoons of it to your food throughout the day. Easy ways include sprinkling it in oatmeal, smoothies or yogurt.

Other sources of ALA include:

Algae oil.

Canola oil.

Chia seeds.

Edamame.

Flaxseed oil.

Soybean oil.

Walnuts.

Talking to a healthcare provider is so important. Your provider knows you and your medical history best. They're prepared to sift through the latest research and tell you what these findings mean for you. They'll also give you individualized guidance on how to get the omega-3s your body needs.

Ref From –

https://my.clevelandclinic.org/health/
articles/17290-omega-3-fatty-acids

https://www.ncbi.nlm.nih.gov/pmc/articles/
PMC7551800/

Troesch, B., Eggersdorfer, M., Laviano, A., Rolland, Y., Smith, A., Warnke, I., Weimann, A., & Calder, P. (2020). Expert Opinion on Benefits of Long-Chain Omega-3 Fatty Acids (DHA and EPA) in Aging and Clinical Nutrition. Nutrients, 12(9), 2555.

Chapter – 33

SIRT6 and Anti-aging

This is a comprehensive review of the biochemical, molecular, cellular, and organismal roles of the histone deacetylase SIRT6. This unique enzyme evolved in eukaryotes to perform multiple critical roles in modulating gene expression, metabolism, DNA repair, and lifespan. In this context, SIRT6 has been demonstrated to play key roles as a tumor suppressor and a critical modulator of metabolic homeostasis.

In recent years, mammalian sirtuins have emerged as critical modulators of multiple biological processes, regulating cellular metabolism, DNA repair, gene expression, and mitochondrial biology. As such, they evolved to play key roles in organismal homeostasis, and defects in these proteins have been linked to a plethora of diseases, including cancer, neurodegeneration, and aging.

Nicotinamide adenine dinucleotide (NAD) was originally identified as an essential coenzyme of energy metabolism. Later studies have revealed that it also plays a key role in various cellular functions related to the aging process. Notably, aging is accompanied by a gradual decline in tissue and cellular NAD+ (oxidized form of NAD) levels in rodents and humans.

This decline in NAD+ levels has been functionally linked to numerous aging-related diseases. SIRTs are NAD+-dependent enzymes that catalyze the post-translational modification (PTM) of various enzymes and signaling factors, including deacetylation (SIRT1-SIRT7), decrotonylation (SIRT3), ADP-ribosylation (SIRT4 and SIRT6), diacylation (SIRT6), and desuccinylation, demalonylation, and deglutarylation (SIRT5).

The decline in NAD+ availability over age contributes to the aging-related decrease in SIRT activities. The functions of SIRT isotypes and their sensitivity to fluctuations in intracellular NAD+ levels are influenced by their different subcellar localizations.

SIRT1 and SIRT2 are found both in the nucleus and cytoplasm in a cell or tissue-dependent context. SIRT3, SIRT4 and SIRT5 are mitochondrial proteins that regulate energy metabolism, and SIRT6 and SIRT7 are mostly nuclear proteins.

Certain SIRT3 subtypes (full-length or truncated form) can also be found in the cytoplasm and nucleus.

SIRTs regulate many longevity-related biological processes such as DNA repair, autophagy, inflammation, protection against oxidative stress, and metabolism, and therefore, they have been linked to a plethora of aging-related diseases, including neurodegeneration, atherosclerosis, chronic kidney diseases, and chronic obstructive pulmonary disease.

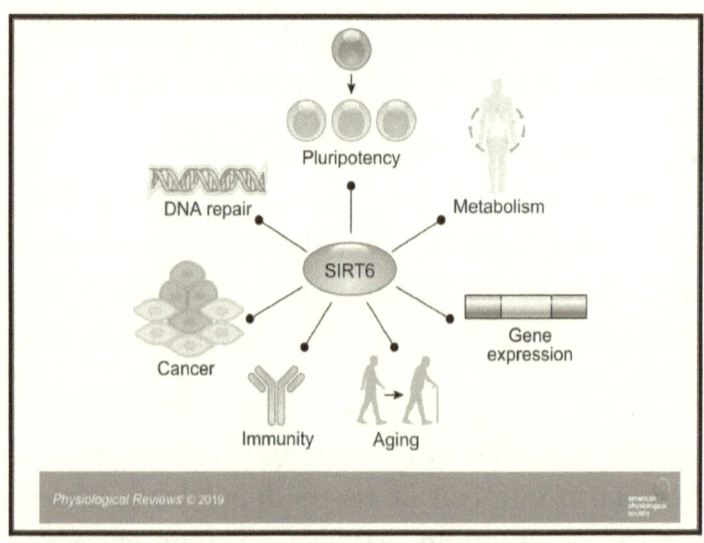

Picture Courtesy From - https://journals.physiology.org/doi/full/10.1152/physrev.00030.2018

SIRT6 is a longevity protein that can inhibit the aging of cells, tissues, organs, and the body by promoting DNA damage repair, maintaining normal chromosome structure, and regulating energy metabolism and the senescence-associated secretory phenotype (SASP). SIRT6 also inhibits immunosenescence, an aspect of aging.

In addition, SIRT6 can regulate the development of inflammation. At present, most evidence reflects the anti-inflammatory effects of SIRT6 *via* inhibiting the production of inflammatory cytokines and promoting the polarization of immune cells to an immunosuppressive phenotype.

SIRT6 is an important protein with anti-aging effects on cells, tissue, organs, and the body. It inhibits aging *via* four main pathways: promotion of DNA damage repair, maintenance of the normal telomere structure of chromosomes, regulation of glucose and NAD+ metabolic balance, and regulation of SASP.

SITR6 regulates inflammatory development and plays a complex role. It inhibits inflammation by promoting M2 macrophage polarization, decreasing the number of lymphocytes, inhibiting

T cell differentiation, and inhibiting the innate immunity response.

SIRT6 is a longevity protein that delays the aging process and participates in maintaining telomere and genome stability. A study at the organism level showed that SIRT6-deficient mice had small body size, with loss of subcutaneous fat, profound lymphopenia, lordokyphosis, and severe metabolic defects 2–3 weeks after birth, and eventually died at approximately 4 weeks of age.

SIRT6 deficiency leads to hyperacetylation of histones at the imprinting control region of developmental repressor H19, which results in severe prenatal developmental delay and death several hours after birth in SIRT6-deficient monkeys.

SIRT6 is a nucleolar chromatin-related protein involved in various DNA damage repair processes. It is related to base excision repair (BER), which has been shown to promote resistance to DNA damage in mouse cells, suppress genomic instability, and promote normal DNA recombination. SIRT6 can also maintain normal chromosome structure.

The stability of chromosomes and telomeres is particularly important for cell anti-aging. SIRT6-deficient cells exhibit abnormal telomere structures. In addition to maintaining genomic stability, as mentioned above, SIRT6 slows the process of aging by regulating glucose homeostasis and the NAD+ metabolic balance. Overexpression of SIRT6 is conducive to a "young state" of blood glucose and gluconeogenesis in aged mice.

Ref From –

https://journals.physiology.org/doi/full/10.1152/physrev.00030.2018

https://www.sciencedirect.com/science/article/pii/S0925443923001813#:~:text=SIRTs%2C%20especially%20SIRT1%20and%20SIRT6,DNA%20damages%2C%20and%20metabolic%20dysfunction.

https://www.ncbi.nlm.nih.gov/pmc/articles/PMC9019339/

My Final Word

These latest research and discoveries in medical science by scientists, biologists, and doctors suggest that **the deficiency of potential vitamins/supplements/molecules** causes a wide spectrum of diseases and impacts the cellular aging process.

Therefore, Vitamins are some of the most potent weapons for fighting anti-aging, but unfortunately, sometimes we can't get the amount we need through diet alone. However, food rich in good nutrition or moderate supplements (under the guidance of a doctor or nutritionist) can fight the symptoms of aging, whether it's skin sagging, cognitive decline, poor physical health, or any other metabolic syndrome.

We want to remind you that **the articles on different supplements in the book are not a substitute for professional advice.**

The author is not a doctor or dietician. However, as a researcher in Medical Astrology, he has done extensive reading over the past few years on Mitochondrial Dysfunction and how it affects the cellular aging of cells, tissues, and organs, which has inspired him to write this subject.

This book presents information strictly for educational purposes. It's tailored for those interested in the latest "health and wellness" research.

The author urges you not to engage in self-diagnosis or treatment based solely on the knowledge acquired from this book. You must consult a medical practitioner for any health concerns, as this book does not claim to provide any medical solutions for any health issues.

The content of this book is for informational purposes only and is not intended to diagnose, treat, cure, or prevent any condition or disease. You understand this book is not intended as a substitute for consultation with a licensed practitioner. Please consult your physician or healthcare specialist regarding the suggestions and recommendations made in this book. The use of this book implies your acceptance of this disclaimer.

This publication is meant as a source of valuable information for the reader; however, it is not meant as a substitute for direct expert assistance. If such assistance is required, the services of a competent professional should be sought.

Courses and Consultation

You will find highest numbers of past predictions of my clients, which have ever been given by any astrologer with 90 percent accuracy in written format on my Website, Facebook profile and FB page (WhatsApp chat transcript).

You can enroll in following courses:

1. **PMD Program** of One Year complete **PROGRAM** of Astrology (combination of Predictive Astrology + Medical Astrology + DNA Astrology Karma Correction).

2. Independent Single course - Advance Predictive Astrology with Timing of Event.

3. Independent Single course – Advance Medical Astrology Course with timing of the disease.

4. Independent Single course - DNA Astrology of Karma Correction with Cutting the Chord.

5. Bhrigu Nandi Nadi

6. **Acharya Upadhi Certification (Final Course of Astrology after PMD Program).**

7. Lal Kitab Recording are available on my website: **Https://www. kaalhasthiastrologer.com**

You can book your consultation directly from the website or you can whatsapp me.

Contact Me

Website

Https://www.kaalhasthiastrologer.com

Facebook

https://m.facebook.com/pages/category/
Astrologist---Psychic/KaalHasthi-Astrologer-Mr-
S-Prakash-569188060151088/

YouTube channel

CosmicKrishna

https://www.youtube.com/channel/
UCFObfsKxwkMV-TRk9fbDI2g

Instagram

https://www.instagram.com/kaalhasthi/

Email ID

kaalhasthi@gmail.com

Telephone

+91 9606873053